A Practical Guide to Vertebrate Mechanics

A thorough understanding of the form, function, and design of animals is essential to any working biologist's knowledge. In the author's view, however, this fast-growing field of study can be made much more exciting and accessible with a hands-on, practical approach. This is the view of *A Practical Guide to Vertebrate Mechanics*.

This text can be considered an engineering book for biologists. The emphasis is on vertebrates, and each topic begins with a discussion of the underlying principles, followed immediately by practical experiments and laboratory exercises. First we begin with a refresher on scaling and measurement. This is followed by three chapters on the mechanical properties of materials: investigating elasticity, the strength of materials, and how things break. This leads our discussion to animal materials – bones, joints, muscles – which serve to illustrate principles of structure and load, lubrication, physiology, metabolism, and stamina. Finally, we put the systems in motion, as we discuss terrestrial locomotion, flight, and swimming.

What sets this book apart from others on functional anatomy is its emphasis on practical work. Many of the experiments are simple to conduct. Detailed instructions for setting up the experiments are given in an appendix, and sample results are included to guide the student.

A Practical Guide to Vertebrate Mechanics will form an important part of undergraduate and beginning graduate courses for zoology, anatomy, biomechanics, and paleontology students.

Christopher McGowan is Professor in the Department of Zoology at the University of Toronto and Curator in the Department of Palaeobiology at the Royal Ontario Museum. His previous books include, *The Raptor and the Lamb: Offense and Defense in the Living World* (1997), *Make Your Own Dinosaur Out of Chicken Bones: Foolproof Instructions for Budding Palaeontologists* (1997), and *Diatoms to Dinosaurs* (1994).

A Practical Guide to Vertebrate Mechanics

CHRISTOPHER McGOWAN

Illustrations by Julian Mulock

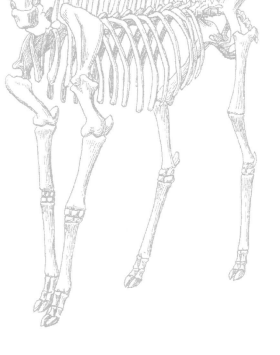

CAMBRIDGE
UNIVERSITY PRESS

PUBLISHED BY THE PRESS SYNDICATE OF THE UNIVERSITY OF CAMBRIDGE
The Pitt Building, Trumpington Street, Cambridge, United Kingdom

CAMBRIDGE UNIVERSITY PRESS
The Edinburgh Building, Cambridge CB2 2RU, United Kingdom   http://www.cup.cam.ac.uk
40 West 20th Street, New York, NY 10011–4211, USA   http://www.cup.org
10 Stamford Road, Oakleigh, Melbourne 3166, Australia

First published 1999

Printed in the United States of America

Typeset in 9.5/13 Stone Serif in QuarkXPress {GH}

*A catalogue record for this book is available from
the British Library*

Library of Congress Cataloging-in-Publication Data
McGowan, Christopher.
   A practical guide to vertebrate mechanics / Christopher McGowan.
      p.   cm.
   Includes bibliographical references and index.
      1. Vertebrates – Morphology.   2. Vertebrates – Physiology.
   3. Animal mechanics. I. Title.
   QL805.M36   1999
   571.3'196 – dc21   98-29462
                                                                 CIP

ISBN 0 521 57194 4 hardback
ISBN 0 521 57673 3 paperback

FOR STUDENTS PAST, PRESENT, AND FUTURE

Contents

Acknowledgments

This book arose from a series of discussions I had with Robin Smith, former Life Sciences Editor with Cambridge University Press (CUP). Robin was enthusiastic and tremendously helpful throughout the formative life of the book, for which I offer my sincere thanks and gratitude. Michael Penn took over as editor following Robin Smith's departure. I thank him, Catherine Felgar, and Jacqueline Mahon, all of CUP, for their various roles in seeing this book to completion. My sincere thanks for the exacting and efficient production staff: Fran Bartlett, Mary Jo Rhodes, and Kathie Kounouklos.

I am extremely fortunate to have had the entire manuscript read by three critical reviewers: Professor Michael J. French, Engineering Department, Lancaster University; Professor Jeffrey J. Thomason, Department of Biomedical Sciences, Ontario Veterinary College, University of Guelph; and one anonymous reader. Dr. James De Laurier, Institute for Aerospace Studies, University of Toronto, read the sections dealing with flight and flying. All four of these busy academics took the time and trouble to carefully review the material, for which I offer my sincere thanks. Their valuable criticisms and corrections have considerably improved the manuscript, for which I am deeply grateful.

Joan Burke, of my own Department of Palaeobiology, the most meticulous and conscientious of readers, carefully proofread the entire manuscript against the page proofs. I thank her for this painstaking task.

Julian Mulock has illustrated almost all of my scientific publications, and several of my previous books, over a period of more than two decades. Working with Julian is always a great joy and I am happy to acknowledge the wonderful and numerous illustrations that grace this work. As usual, he has been unstinting of his time and energy, and I cannot adequately express my gratitude for this. Thank you Julian.

I gratefully acknowledge the Natural Sciences and Engineering Research Council of Canada for supporting my research endeavors over the years.

I am fortunate to work at an institution that still values scholarship, and I am happy to acknowledge the Royal Ontario Museum for providing an environment in which academic goals can be pursued. I am equally happy to acknowledge my long association with the Department of Zoology, University of Toronto, an affiliation I have always treasured. The university has provided me with generations of inquisitive students, who have done so much for my academic growth. My sincere thanks to all these student, to whom this book is warmly dedicated.

Prologue

Prologues are frequently the least read part of any book, but it is important to read this one; otherwise, you will miss the point of the book.

Science, for the most part, is a practical endeavor, and the best way to learn is through a hands-on approach. Unfortunately this is not always the way it is taught, and school science is often reduced to an assimilation of fact and theory where memorizing frequently plays a more important role in learning than understanding. Little wonder that so many youngsters get turned off to science at an early age. My own offspring did. I recall one of my daughters seeking my help with a piece of science homework when she was about eleven years old. She was given some readings on chemistry and was having trouble with some of the chemical equations. The relevant passage in her textbook was illustrated with several test tubes in which the reactions were taking place. I naturally thought the reading was in preparation for conducting the experiments in the lab the following day. Not so. This *was* the practical. The nearest her class was going to get to heating compounds in test tubes to see how they reacted was to read about it. There is no fun in that; but science, to my mind, is meant to be fun, meant to be exciting and challenging and meant to be learned through practical experience. Unfortunately, the trend away from practicals extends into our universities. Many science courses are without practicals, and for those that do have them, the labs are often tacked on as appendages with little integration with the lecture material.

This book has arisen from one of the courses I teach at the University of Toronto. It is essentially an engineering course for biologists where we explore the underlying physical principles of living systems. As I am primarily interested in vertebrates, especially in their functional anatomy, the course is centered on them but not to the total exclusion of invertebrates. Some time is spent analyzing the mechanics of the vertebrate skeleton from the physical properties of its components – bone, tendon, cartilage – to the engineering of its design. Life and movement are then given to the bare bones with a treatment of muscles and exercise physiology. It is here that we learn why reptiles have far less stamina than birds and mammals and how these differences are reflected in their different hunting strategies. This understanding leads to the subject of locomotion, pursued on land, at sea, and in the air.

The format of the course is a one-hour lecture followed by a three-hour practical, specifically on the subject of the lecture. As physical systems are simpler and more easily understood than living ones, they form the starting point for investigations of biological systems. For example, we spend some time pulling pieces of metal apart and breaking things in the Department of Metallurgy and Materials Science as a prelude to assessing the engineering properties of bone. This procedure involves the use of a fairly sophisticated instrument, but most of the other experiments are conducted with far less elaborate equipment, much of it homemade. I had as much fun devising the experiments as my students have had in conducting them, with all of us learning from experience rather than from books. Indeed, there is no recommended text for the course, which is one of the reasons for writing this one. The other and more important reason for this book is to present the subject of functional vertebrate anatomy, an exciting and actively growing field of biology, from a practical perspective.

You can use this book as an adjunct to an existing practical course offered at your own university. If such is not available, perhaps this book will encourage some professors to devise one of their own. To this end I have included some notes to instructors at the end of the book which should help save them time in developing a course. For example, I spent several weeks trying to set up some photoelastic plastic experiments – those transparent models seen illustrated in books showing stress patterns in plastic bridges. None of the books that I could find explained how to set up the models; all they did was make reference to Plexiglas and to polarized light. Plexiglas *can* be used but is not very good, whereas certain other plastics work remarkably well. The resulting stress contours, in dazzling colors, are simplicity itself to produce – once you know how!

Each chapter begins with a discussion of the subject. A series of experiments follows that illustrates and amplifies this material. Ideally you will be conducting experiments like these in your course and can use this section for reinforcement. However, if you are not enrolled in such a course, you can get a feel for the lab work by reading through the description of the experiments. You can then use the sample observations, taken from actual experiments, to calculate your own results. Quite often the results obtained depart from the expected, raising some interesting questions. The likely reasons for these departures are discussed at the end of each experiment. Since some of the experiments involve readily obtainable materials, you may like to conduct them for yourself or to devise similar ones.

Before closing this prologue, I want to say a few words on the subject of *phylogeny*. Phylogeny is the study of relationships among organisms, an area that has made considerable progress during the last quarter century. Part of this progress is attributable to the availability of new techniques for assessing such relationships. One of the most powerful of these tools is DNA analysis, where the triplet codes contained in genes are revealed in the laboratory. DNA sequencing, for example, has revealed that our closest relative among the primates is the chimpanzee rather than any of the other great apes. The objective of phylogenetics is to establish natural groupings of organisms, that is, a group of organisms that evolved from the same common ancestor. Living birds and certain dinosaurs including *Tyrannosaurus* and *Velociraptor* form a natural or *monophyletic* group, the theropods, which evolved from a common ancestor. Descent from a common ancestor is established by looking for specialized or derived features that all members of the group share. Theropods share a number of derived features, including: a foot supported by three long metatarsal bones; three main toes and a short "big toe" that faces backward; a femur that is slightly bowed forward; and a hand with three fingers, each ending in a sharp claw.[1] (Chapter notes begin on page 264.) Incidentally, since birds are theropods, they are properly referred to as dinosaurs. The evolutionary pathways that a group of organisms can follow are in part determined by their phylogeny. Lamnid sharks, for example, as discussed in the final chapter, lack the deeply forked tail of fishes like the tuna. This is because their skeletons are constructed of cartilage rather than of bone, and since cartilage is nowhere near as stiff as bone, their tails cannot be built so slenderly. In the pages that follow, you'll need to bear in mind the constraints placed on organisms by the phylogenetic baggage they carry.

A Matter of Scale

What happens when things get bigger? Suppose you had two apples in your hand, a small crab apple and a regular-sized apple whose diameter was twice as big. How much heavier would the larger apple be? Obviously, not twice as heavy. In fact, the larger one would be eight times heavier. You could verify this for yourself. There are many other objects to choose from besides apples – potatoes, onions, squash, and tomatoes come immediately to mind. If you went into your local supermarket armed with a ruler or measuring tape, you could find a suitable pair of fruit or vegetables where one was twice as large as the other. If you weighed them you would find that the larger one was not exactly eight times heavier – it might be less than eight or more than eight. Why? There are two reasons. First, the larger object is unlikely to be an exact replica of the smaller one. Thus, if photographs were taken of each one and prints were made to the same size, they could not be exactly superimposed. This is the nature of biological specimens. However, they are close enough to having the same shape (technically, having *geometric similitude*) that the larger one will not be far from eight times heavier than the smaller one.

A second, and probably lesser, reason for the discrepancy is measurement error. As the two objects were essentially spherical, you would have measured their diameters, or their circumferences, if using a tape. But it is unlikely that the diameter of, say, an onion, would be the same all around its circumference, because it is not a perfect sphere. To allow for this irregularity you would need to measure several diameters and take an average. However, even if the object had been a perfect sphere, it is unlikely that you would have been able to measure its diameter without error. As you will see later, measurement error always occurs when attempts are made to measure complex shapes, like animals and plants.

There are two sorts of measurement error, one attributable to a lack of accuracy and the other to a lack of precision. Although the terms accuracy and precision tend to be used interchangeably in our everyday language, they have different meanings when it comes to measurements. *Accuracy* is the closeness of a measurement to its true value. *Precision* is the closeness of repeated measurements of the same character. If you had a set of calipers with bent points, you would never be able to measure the length of a skull accurately, simply because your instrument would always give a false reading. However, if you were careful, you would be able to remeasure the skull length and get the same, or almost the same, measurement every time, showing that your measurements were precise. Measuring the length of a skull, or any other linear dimensions of an animal, is not as easy as measuring a regularly shaped object like a cube or cylinder. Consequently it is necessary to define how such measurements are made, as we shall see later. So much for length measurements, but how do we measure how heavy objects are?

The terms *weight* and *mass* are often used interchangeably, but they are not the same. The mass of a body – expressed in such units as grams, pounds, kilograms, tons – is a measure of the quantity of material, and it is invariant of gravity. If a person had a body mass of 70 kg, that person would have the same mass regardless of whether he or she were on Earth, the Moon, or in space. Weight, however, is dependent on gravity, and has the units of *force*. Force is the product of mass and acceleration, in this case the acceleration due to gravity. However, in everyday usage, weight has the same units as mass. A person might therefore be said to weigh 70 kg. But that weight would drop to about 11 kg on the Moon, simply because the Moon's gravity is only about 0.16 that of Earth's. In space a person's weight is zero because there is no gravity. Because of the confusion and ambiguity between the two terms, only mass will be used in this book.

Needless to say, metric units will be used throughout, with the occasional reference to imperial units, as these are universally used in science. However, the metric system is not without its problems, largely because of the many different units that are in use. To resolve these problems an international commission was established in 1960 to recommend which metric units should be used and how they should be abbreviated, and to define some new terms. This system of metric units, universally used in science and adopted by many countries, including Canada, is referred to as SI, from the French *Le Système International d'Unités*. The common SI units used in science to derive such variables as force, *work* (the product of force and distance), and *power* (the rate of doing work), are the meter (abbreviated m), the kilogram (kg), and the second (s). In my undergraduate days the units used to be the centimeter (cm), gram (g), and second.

Let's return for a moment to the relationship between length and mass. The reason why doubling the length increases the mass by eight is because volume, hence mass, increases with the cube of the length. This can be shown with a series of cubes built out of sugar cubes (Fig. 1.1). The cubes have edges of 1, 2, 3, and 4 sugar cubes. Their respective numbers of sugar cubes, hence their volumes and masses, are therefore 1, 8, 27, and 64. The fourth cube, for example, is four times longer than the first, and is therefore predicted to be $4 \times 4 \times 4 = 4^3 = 64$ times heavier, which you know to be true. Comparisons between any other pair of cubes would give similar confirmations of the predicted relationship.

Figure 1.1. A series of cubes, built from sugar cubes.

The cubic relationship between length and mass holds only if the bodies being compared are the same shape, are solid, and have the same density. Comparing apples and pears does not work very well because of the shape differences, although the other two conditions are met (being solid and having the same density). Nor does it work for balloons of different sizes because most of their mass is accounted for by the rubber skin, that is, by area rather than by volume. Nor would it hold for a comparison between a brick and a gold ingot because of the differences in their densities.

The cubic relationship between length and mass can have some remarkable consequences. Consider two people of similar build, one of whom is six feet, six inches tall (1.98 m), the other five feet, two inches (1.57 m). This is not an enormous size difference, but the taller person would be twice as heavy as the other. This is because mass increases with the cube of the increase in size, and $[1.98/1.57]^3 = 2.0$. There are many similar examples lurking on supermarket shelves. I have two similarly shaped plastic containers of bathroom cleaner in front of me. The larger one is only a bottle cap and shoulder taller than the other (24.3 cm compared with 19.7 cm), but it contains twice as much cleaner (500 ml compared with 250 ml). What about different sized animals? Spector (1956) gives some body length and mass data for fishes, including a bonito (*Katsuwonus pelamis*) and a bluefin tuna (*Thunnus thynnus*). Both are members of the same family (Scombridae) and are of similar shape. The body lengths are given as 39 and 116 inches (approximately 1 m and 3 m), so the increase in mass should be by a factor of $[116/39]^3 = 26.3$. The mass of the bonito is given as 39 pounds (17.7 kg), producing a predicted value of 1,025.7 pounds (466.2 kg) for the tuna. The recorded mass is 977 pounds (444.1 kg), which is close. To take a more extreme example, a house mouse (*Mus musculus*) is about 8 cm long (ignoring the tail) and has a mass of about 36 g. An average African elephant (*Loxodonta africana*), which is only roughly similar in shape to a mouse, has a length of about 4 m. The magnification factor here is about 50, suggesting that the elephant would be about 125,000 (50^3) times heavier than the mouse. Multiplying 36 g by 125,000 gives a mass of 4.5 metric tonnes (abbreviated t), which is about the mass of an average elephant. Incidentally, a metric tonne is 1,000 kg, and an imperial ton (note different spelling) is 2,240 pounds, or 1,018 kg, which is virtually the same.

Whereas volumes, hence masses, increase with the cube of the length increase, areas increase only with the square. This can be shown by looking at the sugar cubes again. Consider the surface areas of the four cubes. Each cube has six faces, and the areas of each face are accordingly 1, 4, 9, and 16 for the four cubes (the units are "square sugar units"). The total surface area for the four cubes is therefore 6, 24, 54, and 96, respectively. Comparisons between any pair of cubes confirms the relationship between size increase and area increase. Take, for example, the first and third cubes. The size increase is 3, and the increase in surface areas is 54/6 = 9, which is (size increase)^2.

To summarize, as things get bigger, their volumes increase with the cube of the length increase, whereas their areas increase only with the square of the length difference. As a consequence, areas lag behind volumes, which can be expressed by the area-to-volume ratio. As we shall see, the area-to-volume ratio has universal importance. The values of this ratio for the four sugar cubes are 6/1 = 6, 24/8 = 3, 54/27 = 2, and 96/64 = 1.5. To see how area-to-volume ratio relates to size increase, all we need to do is make some pairwise comparisons between the cubes. For example, the size difference between the first and second cube is 2, whereas the difference in their area-to-volume ratios is 3/6 = 1/2. Similarly, the size difference between the first and third cube is 3, whereas the difference in their area-to-volume ratios is 2/6 = 1/3. For the first and fourth cubes the size difference is 4, and the difference in ratios is 1/4. The pattern is obvious: as things get bigger, the area-to-volume ratio changes according to the reciprocal of the size increase. The universal importance of this relationship, both in living and nonliving systems, cannot be overemphasized. It accounts for an infinite array of phenomena among living organisms, from why a mouse's heart beats about 700 times a minute compared with an elephant's 30, to why the heaviest animal that has ever lived, the blue whale (*Balaenoptera musculus*), is aquatic. Examples from the nonliving world range from why the peas on your plate cool far more rapidly than the baked potato, to why the development of 747-sized aircraft had to await the advent of the jet engine.

One aspect of area-to-volume ratios that I want to focus on concerns load bearing, with all its implications for skeletal support in vertebrates. Before returning to the four cubes to see how their load bearing changes with increasing size, we have to consider how to express load bearing. The simplest way is to assess the pressures acting on the bases of the four cubes. Pressure, or *stress*, the synonymous term that is used in preference here, is defined as force per unit area. The SI unit of force is the *newton* (N), 1 newton being defined as the force required to give a mass of 1 kg an acceleration of 1 m per second per second (1 ms^{-2}). Area is expressed in square meters, therefore stress is expressed in newtons per square meter (N/m^2). The SI unit for stress is the pascal (Pa), 1 Pa being the stress experienced when a force of 1 N acts upon an area of 1 square meter.

What stress do my shoes exert on the ground when I am standing still on both feet? My mass is 77 kg, the acceleration of gravity is 9.8 ms^{-2}, and my shoes have a total area of contact (both soles and both heels) of about 0.028 m^2. The force exerted by my body mass is therefore 77 × 9.8 N, so the stress is $77 \times 9.8/0.028 = 26{,}950$ Pa. As the stresses encountered in biology and in engineering often involve high numbers, it is usual to express results in multiples of pascals, such as the kilopascal, kPa, which is 1,000 Pa. The stress exerted by my body mass is therefore expressed as 26.95 kPa. Atmospheric pressure and blood pressure, still usually expressed in terms of millimeters of mercury, are beginning to be expressed in kPa. Atmospheric pressure is about 100 kPa, and normal blood pressure during systole (the beating of the ventricles) is about 16 kPa. The average pressure of the air inside a car tire is about 200 kPa.

How do the stresses acting on a body change with increasing body length? For the cubes, length is the same as height and width. For simplicity, I shall use their individual mass – the number of sugar cubes in each large cube – to represent force, thereby saving the trouble of multiplying each one by the acceleration of gravity. This shortcut is legitimate here because I am only comparing one cube with another.

The area of the bases of the four cubes of sugar are 1, 4, 9, and 16, respectively, so the stresses acting on their bases are $1/1 = 1$, $8/4 = 2$, $27/9 = 3$, and $64/16 = 4$. Thus stress increases according to the increase in size. The implication of this result is that the stresses exerted on the ground by an animal's feet are proportional to its size – double the length of an animal, and the loading on its feet is doubled. This relationship, of course, assumes geometric similitude. The relationship has important implications for vertebrate skeletons. The body mass of a terrestrial vertebrate is supported by its legs, and the stresses exerted by this mass are ultimately borne by the cross-sectional areas of the leg bones. Given that an elephant is about fifty times longer than a mouse, its bones theoretically have to withstand stresses that are fifty times higher. Bone, like any other material, has finite strength limits, so how should a large animal's skeleton be adapted to deal with the problem? One obvious strategy would be for the limb bones of large animals to be more robust than those of smaller ones. But how much should bones increase in girth in order to maintain the same stresses? That question can be answered by returning to the sugar cubes.

The cubes are geometrically similar, so the stresses exerted on their bases increase in step with the size increase. If you stood on a sugar cube it would probably crumble, so if you built a big enough cube out of them, the stresses acting on the bottom layer of sugar cubes would be high enough to crush them. Suppose that sugar cubes were very much weaker than they are, and that the maximum stress they could withstand before being crushed was equivalent to only a single cube. That would mean that if you stacked one cube on top of another, the one beneath would crumble. So how could you rearrange the four cubes to avoid crushing? Obviously they would have to be arranged into a single layer. The first cube would stay as it was. The second, which contains 8 sugar cubes, could almost be arranged into a perfect square whose edge was 3 cubes long, but there would be one cube short. The total

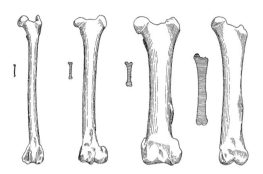

Figure 1.2. Selected femora from animals of increasing body mass. Left to right: dog, deer, hippopotamus, and African elephant. The relative lengths of each bone are shown in silhouette.

area of this irregular shape is 8. If we could crunch it into a perfect square, the edge would be slightly less than 3. In fact, the edge would need to be the square root of 8. This is because when the edge length is multiplied by itself to obtain the area of the square, this area is 8 ($\sqrt{8} \times \sqrt{8} = 8$). Similarly, if the third cube, which contains 27 sugar cubes, were made into a perfect square, its edge would be $\sqrt{27}$. The last cube, comprising 64 sugar cubes, would become a square of edge $\sqrt{64}$. The edge lengths of the four cubes are therefore $\sqrt{1}$, $\sqrt{8}$, $\sqrt{27}$, and $\sqrt{64}$, which can be rewritten as $\sqrt{\text{volume}}$. But volume is length^3, so the edge lengths can be written as $\sqrt{l^3} = l^{3/2} = l^{1.5}$. Therefore, to maintain the same stress, the newly configured sugar shapes have to increase in length, or width, according to $l^{1.5}$. Accordingly, if the stresses on the cross-sectional areas of a vertebrate's leg bones are to remain the same, the diameter of the bones have to scale according to (body length)$^{1.5}$. Do they?

The concept that the diameters of limb bones should scale with their lengths raised to the power of 1.5, to maintain similar bone stresses, is referred to as the theory of elastic similitude. Although some allometric studies supported elastic similitude (McMahon 1975), or showed that it was essentially true (Alexander 1977), they were confined to ungulates (hoofed mammals). When the studies were extended to include a wide range of vertebrates, they found limb bones to be more nearly isometric in their scaling (Alexander et al. 1979). The African elephant, for example, the largest living land animal, has a remarkably slender femur (Fig. 1.2). If large animals do not have relatively thick limb bones, how do they cope with the increased stresses on their bones? By analyzing films of a running elephant (*Loxodonta africana*) and buffalo (*Syncerus caffer*), Alexander and his colleagues (Alexander et al. 1979) concluded that larger animals keep their feet on the ground for relatively longer segments of each stride. They found that the fraction of the time that a given foot is on the ground, the *duty factor*, was 0.49 for the elephant and 0.35 for the buffalo, which was about one-fifth the elephant's mass. This makes good sense because having a longer duty factor reduces the peak stresses acting on the limbs. This observation seemed to resolve the problem of how large animals cope with their large body masses. However, measuring duty factors accurately is a difficult task, requiring a high-speed camera, and

A. A. Biewener (1983b) pointed out that since Alexander and his colleagues did not use one, their measurement errors could be as high as 30 percent. Biewener used a high-speed camera to measure duty factors for animals ranging in size from mice to horses as they ran on a treadmill. He found that the duty factors were all about the same at similar equivalent speeds. This result suggested that large animals do not reduce bone stresses by having longer duty factors. So what is their solution?

In addition to measuring duty factors, Biewener (1983b) measured the angles of the limb bones. Significantly, the angles of the bones decreased with increasing body mass. Thus, larger animals held their limb bones more nearly vertical, thereby reducing the bending stresses acting upon them. Anatomists have long recognized that heavy-bodied animals, like elephants, hold their limb bones straight, with their legs functioning as vertical supporting columns. Such animals are described as being *graviportal*. Biewener's work therefore quantified that which was generally widely known.

You will be able to test for yourself how limb bone proportions change with increasing length in the last experiment of the practicals. However, before turning to these I want to discuss power functions, graphs, and allometry, which are simply ways of representing and analyzing the concepts we have been dealing with here.

If you had a series of cubes of different sizes and plotted a graph of their lengths (along the x axis) against the area of one of their surfaces (along the y axis), you would obtain a concave-shaped curve (Fig. 1.3). Given that x = length and y = area of one face, the equation of the relationship is $y = x^2$. You could plot a similar graph of length against total surface area. This time you would obtain a steeper curve, whose equation is $y = 6x^2$. Equations like these, where one variable changes according to another variable raised to a power, are called *power functions*. The power term, which is 2 in both of these equations, is called the *exponent*. The gradient of these two graphs is obtained by differentiating the equation. In the first equation, $y = x^2$; therefore, $dy/dx = 2x$. The gradient of the second equation, $y = 6x^2$, is $12x$. Since both gradients have an x term, it follows that the gradient is not constant but increases as x increases. This can be shown by drawing tangents to the graphs at various points.

If the logarithms of the data are plotted instead of the raw data, a straight line graph is obtained because the logarithmic transformation converts the power equation into a linear one. For example, when the second equation, $y = 6x^2$, is converted into logarithms it is transformed into the equation log $y = \log 6 + 2 \log x$. This equation gives a straight line graph that cuts the y axis at log 6 (because when log $x = 0$, log $y = \log 6$). The gradient of the graph is 2; that is, the gradient has the same value as the exponent. This result can be shown by differentiating the logarithmic equation. Thus, if log $y = \log 6 + 2$ log x, then $dy/dx = 2$, the other terms disappearing on differentiation. A general form for these power equations is $y = bx^k$, the logarithmic transformation

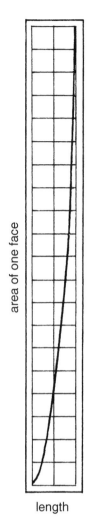

area of one face

length

Figure 1.3. Power functions. The area of one face of a cube plotted against the length of one edge gives a curvilinear graph with an equation of $y = x^2$. The total surface area of a cube plotted against the length of one edge would give an even steeper curve, whose equation is $y = 6x^2$. The gradient of the equation $y = x^2$ is $2x$, and so becomes progressively steeper with increasing values of x.

being $\log y = \log b + k \log x$. If you have a mathematical background, all of this will be tediously obvious. However, if you find it hard going, do not despair. All you need remember are two things: (1) the general form of the power equation, and (2) if logarithmic data are plotted instead of raw data, you get a straight line instead of a curve, the gradient of which is the exponent. The constant, b, can be read from the graph as it intercepts the y axis.

As organisms grow their body proportions change. A familiar example is head growth. Babies, whether human or members of some other vertebrate group, are characterized by their large heads. A human baby's head is about one-third the length of the rest of its body, but as the head and body continue to grow, the head starts lagging behind. By adulthood, the head is only about one-fifth the length of the rest of the body. Such growth, where one part of the body grows at a different rate to the rest, is described as *allometric*, the phenomenon being *allometry* (Greek *alloios*, different; *metron*, measure).

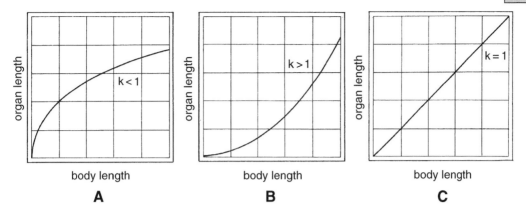

Figure 1.4. Allometric and isometric growth. When the exponent is less than 1, a convex curve is obtained and growth is said to be negatively allometric (A). Positive allometry, when the exponent is greater than 1, gives a concave curve (B). When the exponent equals 1, growth is said to be isometric (C).

Allometric growth is described by the general form of the power equation, $y = bx^k$, where y = organ size, x = body size, k = the exponent, often referred to as the allometric growth constant, and b is a second constant, usually of much less interest. Organ size here might be the diameter of the head, whereas body size is usually taken as body length. When the exponent is less than 1, as in head growth, a convex curve is obtained when organ length (y axis) is plotted against body length (Fig. 1.4A). Such growth is often referred to as negative allometry, because the organ becomes progressively smaller relative to the body length. The exponent for head growth has a value of about 0.7 for a wide variety of vertebrates. When the exponent is greater than 1, the organ becomes progressively larger, relative to the body length, as the animal grows. Growth is said to be positively allometric, and the curve obtained when raw data are plotted is concave (Fig. 1.4B). One of the most noteworthy examples of positive allometry is the claw of the male fiddler crab (*Uca pugnax*). During the early stages of development, both of the male's claws are small, but as growth proceeds, one of the claws becomes progressively larger until, in the adult, it is almost as big as the rest of the body. The strong positive allometry of the male's claw is a sexual character, the giant claw being used to attract females. The male's other claw, and both of the female's claws, grow in step with the rest of the body. Here the exponent is 1 (Fig. 1.4C), and growth is said to be *isometric* (Greek *isos*, equal).

The change in proportion as animals grow larger has been recognized for over a hundred years, but it was not until the first half of the present century that the subject of allometry was treated quantitatively. The central figure in these studies was the English zoologist Julian Huxley, grandson of Thomas Henry Huxley, staunch defender of Darwin. Huxley's book on allometry, published in 1932, had a considerable influence on the biologists of his day and stimulated a plethora of investigations into differential growth. Allometric studies are not confined to growth stages within a species; they include analyses among adults of the same species and adults of different species, too. Huxley (1932), for example, measured antler size against body size for adult red deer (*Cervus elephus*). Antlers are restricted to the male in true deer (cervids), except in the caribou (*Rangifer tarandus*). Like the male fiddler crabs' giant claw, antlers are used in the mating season, primarily in head-to-head ramming contests with rival males. The antlers are shed annually, and each year they grow back bigger than they were before. Huxley found that the antlers had a positive allometry, becoming relatively larger with increasing body length.

Before closing this discussion on allometric growth, I should point out that plotting a double logarithmic graph of growth data does not necessarily result in a straight line. This is because the growth of one part of an animal does not always maintain a constant relationship with the rest of the body. Such cases of nonconformity with Huxley's equation – where the allometric growth constant does not remain the same throughout the animal's growth – are described as *complex allometry*. Conformity with Huxley's equation is referred to as *simple allometry*. Complex allometry is frequently encountered where large size ranges are analyzed. For example, I carried out a study of the relative growth of the rostrum in the swordfish (*Xiphias gladius*). The specimens ranged in size from 6 mm long larvae to 3.7 m adults, and most of the double logarithmic graphs were markedly curvilinear (McGowan 1988). I simplified the problem of analyzing the growth changes by dividing the data into six subsets, based on body length. This division resulted in six linear graphs for each data set.

PRACTICAL

In this lab we will be investigating the relationship between the size of objects and their masses. We will start with nonliving systems, because they are the simplest to investigate, and then look at some biological examples. Whereas physical systems are usually predictable, biological systems are often not, but the departures from the expected results are generally the more interesting.

As you will be weighing and measuring, you are provided with a set of calipers and some scales.

Experiment 1 The relationship between length and mass in cubes.

You are provided with a series of plaster cubes of various sizes. Measure the lengths of their edges, in millimeters, and their masses, in grams. Plot a graph of the edge (*x* axis) against the mass (*y* axis). Plot a second graph, this time with logarithmic data. (Do not use double-logarithmic graph paper because the interpretations are less straightforward.)

Sample Observations:

Length of cube (mm)	Mass (g)
(x)	(y)
10	2
20	16
30	43
40	106
52	160
102	1,250
154	4,580
203	9,910

Discussion of Results The first graph should be a concave curve, showing that the exponent is greater than 1. The logarithmic graph should be a straight line, with a gradient of 3, which is the value of the exponent. The value of the second constant, b, in the allometric growth equation $y = bx^k$, can obtained from this graph. The logarithmic transformation of the previous equation is $\log y = \log b + k \log x$. When $\log x = 0$, $\log y = \log b$; therefore, the value of $\log b$ can be read directly from the graph, where the line intersects the y axis. The value of the intersection is likely to be negative, which requires a little thought when antilogging. Supposing the line intersected the y axis at $\log -2.7$. How do we antilog this value? The way to answer the question is to ask what number would give a logarithm of -2.7. Recall that the logarithm of a number has two parts, the *characteristic*, which is to the left of the decimal point, and the *mantissa,* which is to the right. The characteristic signifies where the decimal place goes. For example, the logarithm of 20.0 is 1.3010, that of 200.0 is 2.3010, and that of 2,000.0 is 3.3010. The characteristic is therefore one less than the number of figures before the decimal point in the original number. The logarithm of 2.0 is 0.3010, that of 0.2 is $\bar{1}$.3010, the minus sign of the characteristic being placed over the top of the 1 rather than in front of it. Similarly, the logarithm of 0.02 is $\bar{2}$.3010, and that of 0.002 is $\bar{3}$.3010. The mantissa is always positive, whereas the characteristic can be positive or negative. To return to the original question: if the intersection of the line with the y axis is at -2.7, how do we antilog this number to obtain the value of b? The logarithm of the unknown number is negative (-2.7), but the mantissa of a logarithm is always positive. Therefore, the value of -2.7 must have been derived from a number whose characteristic was -3 and whose mantissa was 0.3, because the sum of -3 and $+0.3$ is -2.7. Therefore, you look up the antilog of 0.3, which is 1,995. The characteristic of $\bar{3}$ shows that the number begins in the third place to the right of the decimal point, so the number is 0.001995, which rounds off to 0.002. This number is approximately the value obtained for b from the logarithmic graph. Thus the allometric equation becomes $y = 0.002x^3$. We can check to see if this equation makes sense by substituting some values for x and y. For example, in the fourth cube, $x = 40$ and $y = 106$. Therefore, $106 = 0.002(40)^3$. The right-hand side of this equation works out to be 128, which is fairly close to 106.

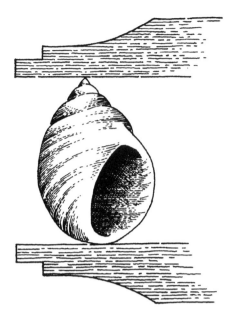

Figure 1.5. Measuring the maximum height of a winkle (*Littorina littorea*) using calipers.

Experiment 2 The relationship between length and mass in various shaped objects.

You are provided with three sets of specimens:

(a) Nuts (7 in the series)
(b) Washers (9 in the series)
(c) Winkles (*Littorina littorea*, 16 in the series)

For each set, record the mass and some measure of the size of each specimen using the calipers and torsion balance provided. Record length in mm and mass in grams, to one decimal place. For the nuts (a), the best measure of size is the distance across the flats; for the washers (b), it is the outside diameter, and for the winkles (c), try measuring the maximum height of the spiral, from the apex of the shell to the outside edge of the aperture (Fig. 1.5). You need to handle the winkles with care, especially the small ones, because it is easy to damage the tips of the spires with the calipers. It is much trickier measuring biological specimens than nuts and bolts – you might want to devise some other measure of size, such as the maximum diameter of the aperture. Whatever dimension you choose to depict size, it is important to define it (a sketch is very useful) so that you can repeat the measurement without a great deal of error.

Sample Observations:

(a) Nuts

Distance across flats (mm)	Mass (g)
7.8	0.87
9.2	1.26

Distance across flats (mm)	Mass (g)
11.0	2.64
19.2	16.28
23.7	33.03
32.7	82.09
37.3	124.30

(b) Washers

Outside diameter (mm)	Mass (g)
9.5	0.48
11.1	0.72
14.2	1.47
19.1	2.51
25.4	5.74
35.2	17.74
44.5	32.09
50.7	45.54
63.3	84.06

(c) Winkles

Height of shell (mm)	Mass (g)
8.1	0.156
8.3	0.149
8.9	0.220
9.0	0.200
11.2	0.366
11.6	0.408
12.0	0.455
14.1	0.710
15.4	0.920
18.4	1.360
20.8	2.240
21.6	2.230
24.7	3.120
25.8	3.590
27.8	4.900
29.2	4.820

Plot two graphs for each set of data:

1. Mass (y axis) against size.
2. Log mass (y axis) against log size. Make sure that your x and y axes are to the same scale, otherwise you will not be able to measure the gradient.

Measure the gradient of the logarithmic graph.

Discussion of Results:

(a) Nuts. Did you discover an exponential relationship between mass and size? You should have found that the first graph was a smooth concave curve, showing

that the relationship is exponential and that the value for *k*, the allometric growth constant, exceeds 1. What value would you predict for *k*? If the objects being measured are geometrically similar (recall, if you took photographs of them and printed them to the same size, they would appear identical), the allometric growth constant should be 3. How similar in shape are the nuts? The larger ones are relatively taller than the smaller ones, therefore relatively heavier. You can check this for yourself by measuring the ratio of height/distance across flats, across the range. This small difference in shape accounts for the small departure from the predicted value of 3 for *k*.

(b) Washers. Did you get an exponential relationship? You should have had a smooth curve for your first graph. From its concave shape you can see that *k* is greater than 1. How far from the predicted value of 3 was your value for *k*? Can you explain any departure from this value? How similar in shape are the washers? You would find it illuminating to measure the thickness of each one, comparing it with the external diameter.

(c) Winkles. You should have gotten an exponential curve for your first graph. Do all the points lie exactly on the graph? You can expect to find some degree of variation in biological specimens, which explains why the points do not all lie on the graph.

Did you expect to get a value of 3 for *k*? The shells are all essentially the same shape, but they are also hollow. Consider the implications of this. If the shells were much thinner walled, would you expect a smaller or larger value for *k*? If the shells were solid, or if they still had the original animals inside, would you expect *k* to be closer to the predicted value of 3? How could you test this hypothesis?

As noted earlier, the cubic relationship between length and mass holds only for similarly shaped objects if they are solid. If most of the mass of the object is attributable to its surface, as in balloons, mass is proportional to length squared and *k* would therefore be 2. Since the winkles have fairly thick walls, the value for *k* is closer to 3 than to 2. However, if the shells were thinner, the value would tend toward 2. If the shells were solid, the value would be close to the predicted value of 3. You could test this hypothesis by filling the shells. You could use water since its density is similar to that of animal tissue, but it is not very practical. Ideally you would select a material whose density is close to that of the shell. Lead shot works well. The density of lead is higher than that of the shell material, but the spaces between the individual spheres reduces this.

Experiment 3   The relationship between length, mass, and stress in cubes.

1. Using the sugar cubes provided, make cubes whose edges are 1, 2, 3, 4, and 5 sugar units long (1 sugar unit = the length or width of one cube). Tabulate your results; namely, the length of the edges of the cubes (in sugar units) and the volume (in cubic sugar units, that is, in number of cubes used). From your background knowledge, you could predict that the volume would increase with the cube of the length, but this simple experiment emphasizes the point.

2. You can also verify the relationship between the mass and cross-sectional sup-

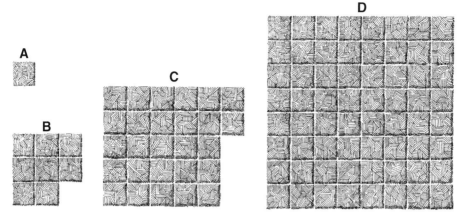

Figure 1.6. Rearranging the sugar cubes of the four cubic models depicted in Figure 1 so that the loading at the base does not exceed unit value. The edges of the new configurations have values of 1 (A), just under 3 (B), just over 5 (C), and 8 (D).

porting area for each of the cubes, that is, the stress acting on their bases, as discussed earlier. This relationship can be expressed as cubic sugar units per square sugar unit. The stress acting on the base of the first cube is obviously 1/1. It is 8/4 for the second cube, 27/9 for the third, 64/16 for the fourth, and so on. Tabulate stress against length of cube for all cubes. Do you see any pattern? The stresses acting on the bases of the first four cubes are 1, 2, 3, and 4. Thus stress increases according to the increase in length.

Suppose the maximum stress that solid sugar could withstand was 1. Obviously, a cubic shape would not be tolerated (except for the single cube). Reorganize each of your cubes into a shape that will not exceed a base pressure of 1, that is, the maximum height of the structure is 1 unit. What are the dimensions of your objects? The first cube stays unchanged. The second cube is dismantled and the individual cubes are rearranged, as far as possible, into a square. You are one sugar cube short of a perfect square, and its edge is something less than 3. If you could squeeze the sugar cubes into a perfect square, its edge would be $\sqrt{8} = 2.83$. When you rearrange the third cube you would find that you could make a square that had a length of 5, with two sugar cubes left over. Again, if this imperfect square could be squeezed into a perfect one, its edge would be a little more than 5 units long, namely $\sqrt{27} = 5.20$. The fourth cube can be made into a perfect square, whose edge is 8 units long, that is, $\sqrt{64}$ (Fig. 1.6). Can you see the pattern? In order to maintain the same pressure, the edge of the base has to increase by $\sqrt{mass}$. Mass = l^3, so $\sqrt{mass} = \sqrt{l^3} = l^{1.5}$. To reiterate what was said earlier, as animals get larger, the diameters of their limb bones would have to increase according to $l^{1.5}$ to maintain the same stress.

Experiment 4   The relationships between mass and length in fishes.

You are provided with a set of fishes of different lengths, all of the same species, preserved in alcohol. Measure the standard body length (mm) and mass (g) of each fish. (Standard body length = tip of snout to base of tail.) As the fishes are wet, you will have to gently dab them damp-dry with paper towels before weighing. Preserved fishes are often eviscerated and saturated with alcohol, so it is not possible to obtain their true body mass. However, since the error in assessing their masses is constant, it does not materially affect the results. Plot a graph of mass (y axis) against length.

Sample Observations (for char, Salvelinus):

Standard length (mm)	Mass (g)
38	0.7
45	1.0
53	1.8
54	2.3
61	3.8
65	3.6
72	5.1
75	5.7
75	6.3
84	7.5
89	8.9
90	10.7
119	23.7
144	43.1
156	55.2
171	55.7
180	78.3

You should obtain a smooth, concave curve when plotting the raw data. Repeat, using logarithmic data. Measure the gradient of the logarithmic graph. Are the results what you would have predicted? You should have found a value for k that is close to 3.0.

Experiment 5   The relationship between body mass and mandibular length in mammals.

You are provided with a set of raccoon skulls and their body mass data. One of the frustrations of working with museum specimens – and most systematic studies are made on such material – is that not all relevant data are likely to have been recorded, especially if the specimens were collected many years ago. Total body lengths (and body masses) are infrequently recorded, therefore some other measure for the size of the individual has to be used. Skull length, or mandibular length, is often used, especially by mammalogists, whereas ornithologists routinely use wing length as a measure of body size. We shall soon see some of the problems with these strategies.

Measure the length of the mandible (mm) and note the body mass (g) for each specimen. It does not matter too much how you place the calipers, provided you are consistent in your measurements. Try measuring the same specimen on several separate occasions, to see how consistent you can be. How close are your measurements? You might like to compare your length measurements with somebody else's. There is often much variation from one person to another, which is why zoologists prefer to do their own measuring.

Sample Observations (for the raccoon, Procyon lotor):

Mandibular length (mm)	Body mass (g)
69.35	252
70.70	376
75.20	346
75.30	444
75.30	468
75.80	504
76.30	534
76.90	566
77.90	365
78.45	470
78.70	548
78.80	524
78.80	680
79.20	624
79.30	500
79.70	416
80.40	716
80.65	714
81.15	696
82.20	706
83.20	588
84.60	742

Plot two graphs, one for raw data, the other for logarithmic data, of mass (y axis) against mandibular length. You should obtain a smooth, concave curve for the raw data. Measure the gradient of the logarithmic graph. Are the results what you might have predicted? You might have predicted a gradient of 3, but the value obtained is considerably higher. Why do you think this is so?

The individuals for which sample data were given ranged from juveniles to adults. Think about the obvious changes in body proportions that young mammals experience when they are growing. The most striking difference between small individuals and adults is that their heads are much larger relative to their bodies. As body mass increases, the head, therefore the mandibular length, becomes progressively smaller. Consequently, when a logarithmic graph of body mass (y axis) is plotted against mandibular length, the gradient is larger than 3. Mandibular length, like skull length, is not the best indicator of body size.

Experiment 6 The relationship between body mass and wing length in birds.

You are provided with a sample of museum "skin" specimens of closely related birds, selected to give a wide size range. Museum specimens are preserved with the wing folded, and it is not possible to stretch the wing out to measure its entire length. Instead, you need to take a standard wing length measurement, which is the distance from the bird's wrist (what appears to be the shoulder region in the skin specimen) to the tip of the longest wing feather (primary feather).

Sample Observations (for tyrant flycatchers):

Species	Body mass (g)	Wing length (mm)
Common tody-flycatcher (*Todirostrum cinereum*)	6.0	41
Northern bentbill (*Oncostoma cinereigulare*)	7.1	48
White-throated spadebill (*Platyrinchus mystaceus*)	8.8	54
Bran-colored flycatcher (*Myiophobus fasciatus*)	9.8	53
Crowned flycatcher (*Onychorhynchus coronatus*)	11.8	71
McConnell's flycatcher (*Pipromorpha macconnelli*)	12.0	67
Vermilion flycatcher (*Pyrocephalus rubinus*)	13.3	72
White-eyed flycatcher (*Tolmomyias sulphurescens*)	14.0	62
Yellow-bellied elaeinia (*Elaenia flavogaster*)	24.1	79
Piratic flycatcher (*Legatus leucophaius*)	25.5	87
Tropical kingbird (*Tyrannus melancholicus*)	39.0	110
Sulphur-bellied flycatcher (*Myiodynastes luteiventris*)	46.0	114
Great kiskadee (*Pitangus sulphuratus*)	68.0	113

Plot a graph of the logarithm of body mass (y axis) against the logarithm of wing length, and measure the gradient. How close is this to the predicted value of 3? You would find that the gradient was less than 3. Can you suggest why?

Consider the function of the wing. The wings have to support the mass of the bird during flight. If wing length were scaled with body mass to give the predicted gradient of 3, linear measurements of the wing would keep in step with increases in the size of the entire bird. However, the wing area would become progressively smaller, relative to body mass, with increasing size. Since lift is generated over the surface of the wing, the decrease in relative wing area would result in a rapid increase in *wing loading* (wing loading = mass of bird/surface area of wings). To help offset this, wings tend to increase in length at a higher rate than with the cube root of the body mass. However, wing loading still increases with body size. For example, the wing loading of a starling is about 5.1 kg per square meter, compared with 8.6 kg per square meter for a pelican.

Why do we not simply measure total body lengths from the specimens and not bother with wing lengths? There are several problems. First, the skins shrink during preparation (the skin has been removed from the body – only the legs, bill, and front portion of the skull remain of the skeleton). There is also considerable variation in the relative lengths of the tails from species to species. An approximate measure of length (i.e., total body length *minus* the tail and minus the bill) can be obtained directly from the specimens and are listed (in the same species order) below as follows: 61, 66, 62, 66, 71, 68, 76, 74, 79, 120, 141, 125, and 165 mm.

Plot a double logarithmic graph of mass (y axis) against these body lengths, and measure the gradient. You should find that the gradient is closer to the predicted value (of 3) than it was for the wing length graph.

One important source of error in assessing size relationships in birds is the extreme variability in their body mass. The mass of a bird varies throughout the day (it is considerably higher after a meal) and throughout the year, being higher at the beginning of the breeding season than at the end. The variations can be as high as 30 percent.

Experiment 7   The relationship between the length and diameter of the femur over a wide size range of vertebrates.

We have seen that the diameters of bones need to increase more rapidly than increases in the length of an animal if the same stresses on their cross-sectional areas is to be maintained. Large animals might therefore be expected to have more robust bones than smaller ones. Body mass and length data are rarely available, and the specimens we are going to examine are no exception. We will therefore use the length of the femur as an index of the length of the animal.

Measure the length and diameter, in millimeters, for each of the femora provided. With the exception of the sauropod, which was added for interest, all of the animals are mammals. To be consistent, take your measurements the same way each time. Measure the length parallel with the longitudinal axis of the shaft, from the head to the distal articular surfaces. To measure the diameter, select the narrowest part of the shaft, which is approximately at its midpoint. Few of the femora have exactly circular cross sections, therefore take an average of the widest and narrowest measurements. Tabulate your results.

Sample Observations:

Species	Common name	Length (mm)	Diameter (mm)
Blarina brevicauda	Short-tailed shrew	11	1.0
Parascalops breweri	Hairy-tailed mole	13	1.0
Scalopus aquaticus	Eastern American mole	16	1.5
Procavia sp.	Hyrax	74	6.0
Taxidea taxus	American badger	90	8.0
Martes pennati	Fisher	90	6.0
Vulpes cinereoargenteus	Gray fox	109	8.0
Felis lybica	African wild cat	127	7.0
Procyon lotor	North American raccoon	128	10.0
Canis mesomelas	Black-backed jackal	142	10.0
Sus sp.	Pig	148	19.0
Lagothrix sp.	Woolly monkey	157	11.0
Gazella thomsonii	Thomson's gazelle	163	15.0
Lynx rufus	Bobcat	174	11.0
Canis latrans	Coyote	178	11.0
Ovis	Sheep	190	22.0
Ursus americanus	Black bear	298	22.0
Cervus elaphus	Red deer	305	32.0
Panthera tigris	Tiger	335	25.0
Bos taurus	Domestic ox	370	40.0
Homo sapiens	Human	420	26.0
Hippopotamus amphibius	Hippo	475	62.0
Loxodonta africana	African elephant	1,320	126.0
Barosaurus sp.	Sauropod	1,370	163.0

Plot a double logarithmic graph of the diameter of the femur (y axis) against the length. Measure the gradient. Is the gradient close to the value of 1.5, as predicted by the theory of elastic similitude? You will find, instead, that your slope is close to 1.0, showing that limb bones are more nearly isometric in their scaling.

Elasticity

Many of the terms used in science and engineering are also used in everyday language, often with quite different meanings, which can cause much confusion. We will encounter a number of such terms in this book, starting with elasticity. If I said that rubber was elastic, which it is, you would have no problem. But if I added that diamond, concrete, glass, and steel were also elastic, which they are, you would probably have some difficulties. In engineering, *elasticity* is used to describe the property where a material returns to its original shape after an external force is removed. Pull on a rubber band and it stretches; let go and it returns to its original length. This is elastic behavior. Pull on a piece of chewing gum and it stretches; let go and it does not return to its original length. This behavior is not elastic. The picture becomes clearer if we think of materials at the molecular level.

The Crystalline Structure of Materials

Most solids, glass being a notable exception, have a crystalline structure. This is also true of liquids and gases when the temperature is below their freezing points.[1] Ice, for example, is a crystal form of water and is formed when water is cooled below its freezing point of 0° C. Gases have much lower freezing points; that of oxygen, for example, is –219° C, and when cooled below this temperature liquid oxygen becomes a solid with a crystalline structure. Crystals have a regular and fixed structure in which the atoms of the material are arranged in space in a certain pattern. There are over a dozen different types of crystal lattice, and each material conforms to one of these patterns (Fig. 2.1). Crystal structures are determined by x-ray crystallography. Here a beam of x-rays is passed through a crystal of the purified material, and the resulting diffraction pattern, caused when the beam is deflected by the atoms in the lattice, is captured on film. Analysis of the film reveals the spacing of the atoms in the crystal lattice, from which their interatomic distances can be deduced. Crick and Watson (1954), for example, used x-ray diffraction data to help arrive at their double helix model for the molecular structure of DNA. The crystal structures of most of the elements have been determined and can be found in several sources (e.g., Pascoe 1978, pp. 4–5).

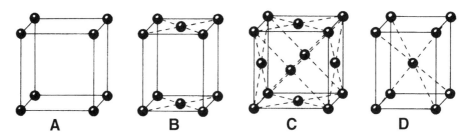

Figure 2.1. Examples of some of the different types of crystal lattice: (A) cubic, simple; (B) orthorhombic, base-centered; (C) cubic, face-centered; (D) tetragonal, body-centered.

For any given crystal, the interatomic distances are constant and are in the order of $2–5 \times 10^{-10}$ m, or 0.2–0.5 nm (nm, nanometers, 10^{-9} m). If a sufficiently large compressive force were applied to a crystal, it would reduce the interatomic distance. However, the atoms would resist the force, and the closer they were forced together, the greater would be their repulsive force. For example, if a force of 4 N were required to reduce the interatomic distances by 1 percent, a force of 8 N would be required to reduced them by 2 percent. Thus, the force of repulsion is proportional to the change in interatomic distance. Similarly, if a tensile (pulling) force were applied to a crystal, it would increase the interatomic distance. But the atoms would resist the force and pull back, and the force of attraction between the atoms would be directly proportional to the increase in their interatomic distances. As soon as the compressive or tensile forces are removed, the atoms immediately return to their original distances apart, restoring the crystal to its original "resting" condition. There is therefore no permanent change in interatomic distances, and such behavior is elastic.

Stress, Strain, and Young's Modulus

It is very difficult to measure the changes in interatomic distances when crystals are subjected to compressive or tensile forces, because they are so small. However, when a test sample of the material is compressed or stretched, its change in length can be measured. This change represents the summation of changes in billions of interatomic distances. Similarly, when passengers standing in a crowded bus are asked to squeeze closer together, their interpersonal distances hardly change, but the accumulated space saving allows several more people to board. As the change in length of a test piece represents the summation of changes in interatomic distances, and as the change in interatomic distance is proportional to the force applied, it follows that the change in length of a test piece is proportional to the force applied.

The change in length of a test piece is expressed by the following ratio: change in length/original length, which is referred to as *strain*. The force applied to a test piece, whether compressive or tensile, is expressed as force/area (cross-sectional area of test piece) and is referred to as *stress.* Since strain ∝ stress, stress/strain is constant. This constant is called *Young's modulus,* after Thomas Young (1773–1829), and is abbreviated *E.* This means that if you plot a graph of stress against strain, you get a straight line. What are the units of Young's modulus? Stress has the units of newtons per square meter, or pascals (Pa). Strain is change in length divided by original length, that is, length/length, and therefore has no units. Young's modulus therefore has the same units as stress (Pa).

Young's modulus of elasticity, to give it its full title, quantifies the *stiffness* of a material. Supposing we wanted to test the stiffness of rubber. We could do this very simply by cutting a rubber band, to give a single strand of rubber, clamping one end in a retort stand, and attaching a weight pan to the other. The latter could be made out of a postcard and Scotch tape. All we would need then is a ruler to measure the extension and some weights to stretch the band. A fairly small weight (stress) would produce a fairly large extension (strain), so stress/strain would be a fairly low number. When we repeat the experiment with a piece of steel wire in the practicals, we have to use a far more elaborate setup because it takes a considerable force to stretch the wire by a very small amount. A large stress would therefore produce a very small strain, so when we calculated *E* (stress/strain), the result would be considerably higher than it was for the rubber. Young's modulus for rubber is about 7×10^6 Pa, which can be written as 7 MPa (megapascals, 10^6), compared with about 200,000 MPa for steel. Steel is therefore far stiffer than rubber, but it is not as stiff as diamond, which has a modulus of 1,200,000 MPa (or 1,200 gigapascals, abbreviated GPa, giga being 10^9). Diamond is obviously a very stiff material, whereas rubber is not very stiff at all; indeed, rubber is the very antithesis of stiffness. Stiffness is a term something like speed. In discussing the speed of various animals we would say that a lion is very speedy, a cheetah is even speedier, and a tortoise is not very speedy at all. However, instead of saying a tortoise is not very speedy, it is simpler to say it is slow. Similarly, rubber may be referred to as being the converse of stiff, namely *compliant,* or flexible.

As you can see, large numbers are involved when calculating *E,* which is mainly because the stress term is a large number. We can show this fact by a worked example. Suppose we wanted to calculate the tensile stress in a steel wire when a 1 kg weight was attached to one end. The force (mass × acceleration) acting on the wire is 1×9.81 Pa. The cross-sectional area of the wire is πr^2, which we need to express in square meters. If the wire had a diameter of 4 mm, the cross-sectional area would be $\pi(0.002)^2 = \pi(0.000004)$ m^2. The stress in the wire is therefore $9.81/\pi(0.000004) = 784,800$ Pa, or 784.8 kPa. When you convert millimeters into meters, it is very easy to make a mistake in the number of zeros placed after the decimal point. Similar mistakes can also be made when squaring these small numbers, so you must take great care when doing the calculations. The situation is exacerbated by the fact that very small numbers are being divided into much larger numbers.

Table 2.1 Young's Modulus for Some Common Materials

Material	E (GPa)
rubber	0.007
polycarbonate (plastic)	2.4
polystyrene (plastic)	2.7–4.2
oak	11
birch	16
bone	20
glass	71
gold	78
copper	130
steel	210
diamond	1,200

Data mainly from Kay and Laby 1995. Values given in gigapascals (Pa $\times$ 10^9).

Values for Young's modulus for some familiar materials are given in Table 2.1. When you see values for E in books you can assume that they were measured in tension, unless stated otherwise. Many materials have the same value for Young's modulus in tension and in compression, but most noncrystalline materials do not. Plexiglas (Perspex), for example, is much stiffer in tension than in compression.

The importance of a material's stiffness should not be underestimated, and it is a common mistake to pay far more attention to its strength. That is not to say that strength is not an important property, too, only that both properties have to be taken into account. An example from my youth will serve to illustrate the point. There were fairly few cars on English roads during the fifties, and most of them were either prohibitively expensive, old and decrepit, or both. One solution to the problem was to buy a cheap car that was in good running order, strip off the rusty old body, and replace it with a fiberglass one. A number of small factories got into the fiberglass laminating business, and there was a range of sporty looking bodies to choose from. The relatively new composite material was strong and light and would never rust. But there was one disadvantage over steel that everyone seemed to overlook – it was not very stiff. As a consequence, doors often did not shut properly where the body had sagged, and the same thing happened for the hood and the trunk.

I said earlier that if a graph of stress is plotted against strain, a straight line is obtained because stress/strain, that is, Young's modulus, is constant. Such elastic behavior is described as being linear, and is typical of crystalline materials (Fig. 2.2A). However, as the stress continues to increase, a point is often reached when the graph departs from linearity (Fig. 2.2B). This point is described as the *limit of proportionality*. Although some noncrystalline solids, such as glass and Plexiglas, also have a linear elasticity, most behave nonlinearly and therefore

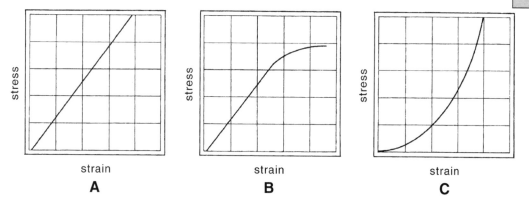

Figure 2.2. Linear and nonlinear elasticity. The elastic behavior of crystalline materials is typically linear (A). However, as stress increases, a point is often reached beyond which elasticity becomes nonlinear (B). Most noncrystalline materials behave with a nonlinear elasticity (C).

have a curved stress/strain graph (Fig. 2.2C). Materials whose elasticity is nonlinear obviously do not have a constant Young's modulus. If you wanted to describe their stiffness, you would have to specify the stress range over which you assessed the modulus. An extreme case of nonlinear elasticity is exemplified by rubber.

Most materials have maximum strains in the order of 0.005 before they break or fail. This means that if you pulled hard enough on a meter-long sample of most materials, it could be stretched by as much as 5 mm before it failed. This strain is only temporary, and the material reverts to its original length when the stress is removed. However, as we will see later, materials often become permanently strained prior to failing.

Rubber, in contrast to most other materials, has a maximum strain of several hundred percent. You can check this fact for yourself by pulling on an elastic band and seeing by how many times longer than its original length you can stretch it. If you conducted a proper experiment, loading the band incrementally and measuring stress and strain until you approached the breaking point of the rubber, you would obtain a sigmoidal stress/strain graph (Fig. 2.3). We can visualize rubber as having a long folded molecule like a concertina. Small stresses cause large strains as the molecule becomes stretched out. Eventually a point is reached when the molecule is fully stretched, whereupon large stresses are needed to produce small strains by increasing the interatomic distances. The rubber breaks some time after this point is reached. Rubber, and similar materials with remarkably large strains, are referred to as *elastomers*. These elastomers include the vertebrate protein *elastin*, found in arterial walls, skin, and certain ligaments, and the insect protein *resilin*, found in wing joints, the ovipositor of certain insects like the locust, and the jumping legs of fleas. Elastomers also behave somewhat similarly during compression. Initially,

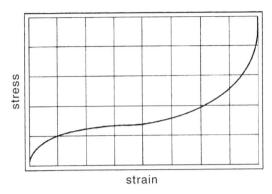

Figure 2.3. Rubber, and other elastomers, exhibit extensive strains and have a sigmoidal stress/strain graph.

small stresses result in large strains as the parts of the molecule become more closely packed together. Eventually the molecule is tightly packed, and large stresses are then required to produce small strains by decreasing interatomic distances. The material fails (breaks) sometime after this point is reached.

Experiencing Elasticity

All solids are elastic to some extent, but we seldom experience that elasticity, except for certain materials like elastomers. String, for example, probably does not immediately come to mind when you think of elasticity. Try holding a short length of string between your hands and pulling hard. Do you feel any "give"? Probably not. Now tie one end of a ball of string to something substantial, like a doorknob, and let out four or five meters. When you try pulling this time, you will be surprised at how springy the string feels. By using a long length of string you have increased the number of interatomic bonds so much that the accumulative effect of billions of infinitesimally small movements are large enough to be felt. Similar results can be obtained for steel wire. However, you would need to use a much longer length because Young's modulus is very much higher for steel than for string.

A less direct way of experiencing elasticity in solids is to try flexing a length of the material. For example, if you bend a saw blade, the steel on the outside of the curve will be stretched slightly, while that on the inside will be compressed by a comparable amount, enabling the blade to be temporarily bent. As soon as you release the blade it springs back straight again, showing that its behavior is elastic. You can repeat the same procedure with a glass rod, to demonstrate the elasticity of glass. (Be very careful if you try this because of the hazards of flying glass if the rod should break. To be safe, wrap the rod in several layers of plastic wrap first, and wear gloves and protective goggles.) Similarly, tall concrete buildings sway slightly in the wind. The top of Toronto's 533 m tall CN-Tower, for example, moves almost half a meter from its center during a strong wind. And many of us have probably experienced how the floor of a shopping mall sometimes vibrates.

Figure 2.4. Loading a length of wire by adding a weight to the pan would produce a very small extension (A). However, if the wire were wound into a spring, considerably more extension would be produced (B).

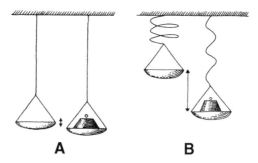

A   **B**

The Distinction between Materials and Structures

If you secured a length of wire to the ceiling and attached a suitable weight pan to the other end, you could measure the extensions produced by the addition of various loads. A graph of extension plotted against load would give a straight line, showing that extension is proportional to load. Young's modulus could then be deduced by calculating the strain for a given stress. The increments in length in the wire for each weight added would be very small (Fig. 2.4A) and could only be measured using a vernier scale, as will be described in detail later. If you disconnected the wire and coiled it tightly around a rod of suitable diameter, you could make it into a spring, provided it was made of mild steel (Fig. 2.4B). Repeating the experiment with the spring would produce similar results as before. Thus the extension would be directly proportional to the load, and a graph of extension plotted against load would be a straight line. However, the extension produced by a given load would now be considerably larger because a structure is being stretched rather than a material. We can attribute the large extension to increasing the distances between the coils in the spring rather than to increasing the interatomic distances in the steel. It would therefore be quite inappropriate to attempt to deduce the strain obtained for a given stress and calculate Young's modulus, because this would give the elastic modulus for a structure rather than for a material. Remember, then, that Young's modulus applies only to materials, never to structures. Incidentally, the linear relationship between extension and load is known as *Hooke's Law,* after Robert Hooke (1635–1702), who first described it. Since the extension of a spring varies with the load, we have a simple and convenient means of weighing things. All that is needed is a spring of the appropriate size and thickness and a scale over which an attached pointer moves (Fig. 2.5). If a weight of 100 g produced an extension of, say, 5 mm, 600 g would produce an extension of 30 mm. The simple weighing device is called a spring balance.

Stiffness in Tension and in Compression

The stiffness of a material is unlikely to be exactly the same in tension as it is in compression. Some materials, like wood, are stiffer when loaded in tension

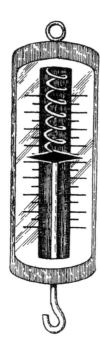

Figure 2.5. A spring balance.

than in compression. Others, like rubber, are stiffer in compression than in tension. We will see in the next chapter that the same also applies to strength, with some materials being much stronger loaded one way than the other.

The Limits to Elasticity

So far we have considered only the temporary changes in length of a material – its elastic behavior. But what happens if stress continues to be applied to a material after it has reached its elastic limit? *Elastic limit* is defined as the limit beyond which the material no longer returns to its original length after the stress is removed. A simple way of demonstrating elastic limit is with one of those plastic supermarket bags. Grip the bag firmly between your two hands, so as to hold the side with the writing toward you. If you pull gently several times, releasing the tension after each pull, you will feel the material stretching. Its behavior is elastic. Now pull again, but this time keep on pulling until the writing becomes distorted, releasing the tension before the material fails (breaks). The material no longer returns to its original length and has been permanently stretched out of shape. The permanent increase in length of the material is referred to as *permanent strain*. The behavior of the material while it is being strained beyond its elastic limit is described as *plastic*, and the change in appearance is described as *plastic deformation*. Materials that undergo extensive plastic deformations are described as being *ductile*. Toffee is ductile, so is modeling clay, chewing gum, the material used in plastic supermarket bags, hot glass, and most metals. Other materials

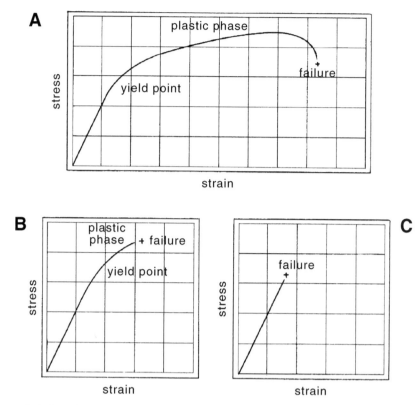

Figure 2.6. Stress/strain graphs for ductile and nonductile materials. Ductile materials (A) have an extensive plastic phase. Nonductile materials have only a limited plastic phase (B) or essentially no plastic phase (C).

exhibit very little or no plastic deformation and are described as being non-ductile. Many nonductile materials are *brittle,* in that they break easily, as when they are dropped or struck with something hard like a hammer. We will see in Chapter 4 that brittle materials are those in which cracks are easily propagated. Toffee is unusual in being ductile and brittle. If you pulled on a piece of toffee it would stretch by a considerable amount. But if instead you hit it with a hammer, it would shatter. Nonductile materials include glass, ceramics (stone, brick, concrete, pottery), cast iron, bone, teeth, and good pastry. The distinction between ductile and nonductile materials is usually readily seen when their stress/strain graphs are compared (Fig. 2.6). When ductile materials are subjected to tensile stress there is an extensive plastic phase before the material fails (Fig. 2.6A). The beginning of this plastic phase, which is marked by a sudden increase in strain, is described as the *yield point.* Nonductile materials (Fig. 2.6B, C), in contrast, fail soon after exceeding their elastic limit. That is not to say that they do not have a plastic phase, only that it is not extensive.

The differences in behavior between ductile and nonductile materials is a

Figure 2.7. Necking occurs in ductile materials, such as soldering wire.

result of differences in the nature of their atomic bonds. The metallic bonds of metals permit the layers of atoms in a crystal to slip or *shear* past each other. The slippage that occurs during the plastic deformation results in a narrowing of the test piece, described as *necking*. The covalent bonds of nonductile materials do not permit slippage. Instead, the interatomic bonds just break, resulting in the sudden failure of the material. You can see necking yourself with a simple experiment. If you take a piece of soldering wire, which is very ductile, and pull it apart, you will see that the two ends have become narrowed (Fig. 2.7).

We see examples of the differences in behavior between ductile and nonductile materials in our everyday life. Windows and crockery are never bent during mishaps, they are broken instead. Cars, in contrast, get bent in accidents, not broken. However, my wife told me of a collision she witnessed where one of the cars "fell apart." From her description of the broken car it was obviously a Corvette, which has a fiberglass body. Fiberglass is a brittle material and therefore breaks rather than bends, which is one of the reasons it is not commonly used for car bodies.

When a brittle material breaks, the broken pieces can be fitted together again exactly. This is because there has been no permanent strain, so there is no distortion in the component pieces. Many of us learned this at an early age, when making surreptitious repairs on broken crockery. Ductile materials, in contrast, cannot be fitted together exactly after they have failed because of the distortion undergone during the plastic phase, immediately preceding failure. You can check this phenomenon for yourself with a simple experiment. Take a small "Post-it note" sized piece of paper and pull it apart. You may have to cheat a little and give it a small twist when you pull, to encourage it to tear apart. Now try matching up the broken edges. You will find that the edges agree very well. This is because paper behaves as a brittle material and does not undergo a plastic phase, so there is no permanent strain. You might not think of rubber as being a brittle material, and it is not brittle in the sense of breaking very easily, as when dropped from a height or struck with a hammer. However, when it breaks, it does so suddenly, without passing through a plastic phase, so there is no permanent strain. Again, you can check this statement with a simple experiment. Take a rubber band, the wider the better, and pull it apart until it breaks. (As a precaution against a painful sting, wear gloves or try using your feet.) If you try matching up the broken ends, you will find they fit together perfectly. As a contrast, try pulling a piece of ductile material apart. You could use a strip of plastic cut from a supermarket bag. Alternatively, if you have a broken tape cassette, video or audio, you could use a piece of that. When these materials are pulled apart they pass through an extensive plastic phase, during which there is extensive plastic deformation. If you tried lining up the corresponding broken edges you would find they did not match at all well. Jell-O is another material that you would not normally think of as being brittle. However, if you try flexing a piece of Jell-O until it breaks, you will find that it does so catastrophically, without going through a plastic phase. If you join up the broken edges afterward you will find that they match perfectly. Next time you are eating Jell-O in company, try convincing your companions that it is a brittle material. I guarantee you will get a very skeptical response!

Before concluding this chapter, I want to say something more about shear. As mentioned earlier, shear has to do with slipping. If you gave a sideways push to a neatly stacked pack of cards, the individual cards would slip relative to one another, and the rectangular pack would become a parallelogram (Fig. 2.8A). Imagine replacing the pack of cards with a block of foam rubber that was glued to the table (Fig. 2.8B). Applying a lateral force, F, would have a similar result, each layer in the rubber being displaced relative to every other. For example, the layer represented by plane *ABCD* was originally in line with plane *EFGH*, but it has now been displaced through the angle theta – similarly for plane *WXYZ*. The angular displacement, theta, is defined as the *shear strain*, usually expressed in radians. When this displacement is small, it is approximately equal to the lateral displacement, *d*, divided by the height of the block, *h*. Shear strain is comparable to tensile strain (Fig. 2.8C; *d/l*). Similarly, there is a variable called *shear stress*, which is given by the shearing force, F, divided by the cross-sectional area of the sample (*ABCD* in Fig. 2.8B). Shear stress divided by shear strain is defined as the *shear modulus*. The shear modulus of a material is a measure of how readily it is deformed by shear stresses, just as Young's modulus is a measure of how readily a material is deformed by tensile or compressive forces. If a block of wood of similar dimensions to the foam rubber block were glued to the table and similar shearing force was applied, the lateral displacement would be considerably smaller. Indeed, no matter how hard you pushed against one side, the lateral displacement would be so infinitesimally small that you would not see any

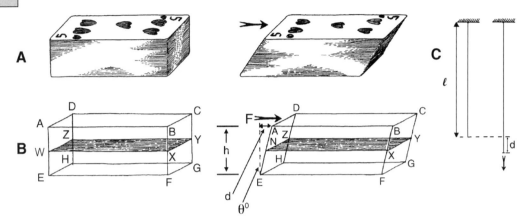

Figure 2.8. Shearing in a deck of cards (A) and in a block of foam rubber (B). Shear strain is comparable to tensile strain, as when a piece of wire is stretched (C).

change in its shape. This is because wood has a much higher shear modulus than foam rubber. As we will see in the next chapter, things can be broken in shearing just as they can in tension and in compression.

PRACTICAL

Experiment 1   The determination of Young's modulus for a piece of wire.

You are supplied with a length of piano wire that has been anchored to the ceiling, a weight carrier with some weights, and a vernier scale, set up as shown (Fig. 2.9). When weights are added to the carrier, one arm of the wire is stretched, and you can measure the amount of extension for a given load from the vernier scale.

Start by noting the reading on the vernier scale with no weights on the carrier. Then load the carrier, one weight at a time, noting the new length. Tabulate the increase in length of the wire and the loads that caused this extension.

Sample Observations:

Load (kg)	Extension (mm)
0.0	0.0
0.5	0.5
1.0	0.9
1.5	1.3
2.0	1.8
2.5	2.3
3.0	2.8
3.5	3.3

Figure 2.9. A simple apparatus for measuring Young's modulus for a length of wire.

The original length of the wire is 2,653 mm, and the diameter is 0.5 mm. Plot a graph of load, in kilograms (y axis), against extension (x axis). Calculate Young's modulus, E.

You could calculate E for each increment, since the stress/strain relationship is linear, but for greater experimental accuracy, you should calculate it for the maximum extension.

$$E = \frac{\text{stress}}{\text{strain}} = \frac{\text{force per unit area}}{\text{extension/original length}} \quad \text{Pa}$$

The graph you obtain is not a graph of stress against strain but of load against extension. To convert *load* into *stress* you would have to multiply by 9.81, then divide by the cross-sectional area of the wire. To convert *extension* into *strain* you would have to divide by the original length of the wire. Since these operations would change each value by a constant amount, the *shape* of the graph would remain unchanged. Therefore, if the original load/extension graph was a straight line – or a concave or convex curve – the stress/strain graph would be the same shape. However, the gradient of the graph, and its position relative to the axes, would change. If you need convincing, just try plotting a stress/strain graph and a load/extension graph and compare the two.

What value did you obtain for Young's modulus? Kaye and Laby (1995) give values in the vicinity of 200 GPa (200×10^9 Pa), depending on the type of steel. From the data supplied, you should have obtained a value of 140.8 GPa.

There are more convenient ways of measuring Young's modulus than suspending wires from ceilings. You will see in the next chapter how mechanical testing machines, like the Instron, can be used to determine E, as well as to test samples to destruction.

Measuring the Strengths of Materials

The strength of a material can be measured in tension or in compression and is referred to as tensile strength and compressive strength, respectively. Strength can also be measured in shear, referred to as *shear strength,* but the experimental procedure is less straightforward, and we will not be concerned with it any further. Strength may be defined as the stress that has to be applied to a piece of material in order to break it. For practical reasons that will be explained later, tensile testing is more often carried out than compressive testing, and there are therefore more data for tensile strength in the literature than for compressive strength. *Tensile strength* may be defined as the maximum tensile force that a piece of material can withstand prior to breaking, divided by its original cross-sectional area. It is important to refer to the original cross-sectional area because this area may decrease as the material is being stretched.

Suppose we wanted to determine the tensile strength of a brass wire. We could estimate the tensile strength by suspending the wire from a fixed point, perhaps a ceiling bracket, and adding weights to the other end until the addition of one more weight just caused the wire to break. Supposing it took 32.45 kg to break a 2 mm diameter wire. The tensile strength of the brass would therefore be $32.45 \times 9.81/\pi (0.001)^2$ Pa = 1.013×10^8 Pa, or 101.3 MPa.

If we wanted to estimate the compressive strength of, say, a sugar cube, we could do so by placing the cube on a flat surface and adding weights until it crumbled. If it took 3.144 kg to break a cube whose edges were 15 mm long, its compressive strength would be: $3.144 \times 9.81/(0.015)^2$ Pa = 13.71×10^4 Pa. We saw in the last chapter that some materials have different values for Young's modulus when they are loaded in compression rather than in tension, and the same is also true for strength. Ceramics, for example, are considerably stronger in compression than in tension, which dictates how they are used in buildings. Indeed, it was the inherent weakness of stone and brick when loaded in tension that was the driving force behind the evolution of new architectural designs. But to understand this weakness we need to know a few basics of building construction.

Beams, Columns, and Arches

The Greeks constructed their large buildings by supporting stone beams, called architraves, on columns (beams are horizontal, columns are vertical).

Figure 3.1. Beams and arches. When a beam is supported at either end, it tends to sag (A); the top surface is loaded in compression, the bottom surface in tension. Stone is weak in tension, limiting the distance between beams (architraves) in ancient buildings (B). The arch is a device for loading stone in compression (C).

When beams are supported at either end, they tend to sag in the middle. The molecules on the top surface are therefore pushed more tightly together, that is, they are under compression (Fig. 3.1A). Those on the bottom surface, in contrast, are pulled farther apart and are therefore in tension. As the distance between the supports of a beam increase, it sags more, increasing the compressive and tensile forces. Because stone is fairly weak in tension, the Greeks were limited in the distance between supports, and their columns are only about as far apart as you can reach with your outstretched arms (Fig. 3.1B). This limitation put a large constraint on building design, because it meant that there could be no wide unsupported areas.

The solution to the problem was to use the arch, a device that loads the individual components, whether dressed stones or bricks, in compression (Fig. 3.1C). Each stone in the arch is tapered, so that they form an arch when fitted together. Since the top and bottom surfaces of the stones are kept in compression, there is no fear of their breaking, and the distance between the supports can be extended considerably. The limitation placed on the building design is no longer determined by the tensile weakness of the stone but rather by the skills of the structural engineer. Steeply curved arches are easier to design and build than flatter ones, and there is a limit to how flat an arch can be before the individual blocks are no longer pressed sufficiently tightly together to keep them locked in.

Isambard Kingdom Brunel (1806–1859) was probably the greatest engineer in Victoria's England. His triumphs included the Great Western Railway and the Clifton suspension bridge across the Avon gorge. He won the latter contract when he was only twenty-two years old, which might explain why he had his critics. During construction of the Great Western Railway, Brunel designed a bridge to cross the River Thames at Maidenhead. The twin arches were remarkably flat, and their 40 m (128 feet) spans may have been the largest ever constructed in brick. His critics predicted that the bridge would collapse. When the wooden supports were removed from beneath one of the arches, it seemed they may have been right because the lowest courses of bricks had separated from the rest for a distance of about 4 m (12 feet) on either side of the crown of the arch. This separation occurred because the supports had been removed before the mortar had set properly, and the contractor admitted his liability and repaired the damage. This time Brunel instructed that the supports be left in place until the spring, but a violent storm blew them away in the autumn of 1839. Fortunately, there was no problem this time, and the bridge has carried railway traffic ever since.

The problem of the tensile weakness of ceramics has been overcome by combining them with materials that are strong in tension. The poured concrete floors of modern buildings, for example, are reinforced with steel rods, and steel bars are often used as lintels over door and window openings.

Isotropy and Anisotropy

If a cube of glass or concrete were compressed in a testing machine, the stiffness and strength would be the same regardless of the orientation of the cube. Similarly, if a piece of steel or Plexiglas were stretched, its stiffness and strength would be the same whether the tension was applied lengthwise or widthwise to the test piece. Materials whose properties are the same, regardless of the direction of loading, are referred to as being *isotropic* (Greek *isos,* equal; *tropos,* change in manner). Many biological materials, in contrast, are not the same in all directions. Wood, for example, as every carpenter knows, is stronger and stiffer when loaded parallel with the grain than at right angles to it. Bone is also stronger and stiffer when loaded parallel to the grain than against it. Material whose properties vary with the direction of loading are referred to as being *anisotropic*. The biological significance of the anisotropy of bone will be dealt with in Chapter 5.

The Mechanical Testing of Materials

A Swiss civil engineering friend, a concrete specialist by training, visited Toronto soon after completion of the CN-Tower. This slender column of concrete, which dominates the Toronto skyline, was the tallest freestanding building in the world. "So what do you think?" I asked, interested in his professional opinion. "They could have built it a lot more slender," he replied. "The column doesn't need to be as thick as that."

Civil engineers do not want their buildings to collapse, but they do not want to overbuild them, either, because that incurs unnecessary costs for additional materials. As a compromise, they build them with a safety factor, meaning that they build them stronger than they need to be for normal conditions. The extra strength is added to take care of unforeseen eventualities,

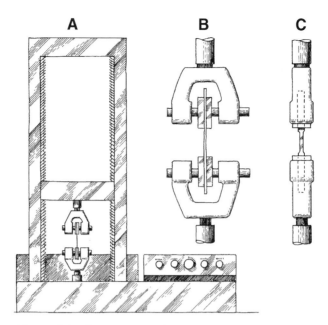

Figure 3.2. The Instron machine, used for the mechanical testing of materials: (A) the complete machine; (B) close-up of the jaws from the front; (C) view from the side, showing how the test piece is gripped.

like abnormally high winds or perhaps the impact of a light aircraft. An engineer's ability to determine how thick to pour concrete, or what gauge of steel cable to use for a suspension bridge, is based, among other things, on his or her knowledge of the mechanical properties of these materials. Two fundamental properties concern us here, stiffness and strength, and I want to describe how they are determined experimentally.

There are several different types of instruments available for mechanical testing, including some sophisticated hydraulic machines, but I will describe one of the older models, of the brand name Instron. The reason for doing so is because of its simplicity, my familiarity with its operation, and its general accessibility in engineering and materials science departments at universities.

The Instron is built as a machine for pulling things apart, but it can be modified for compressive testing, too. It comprises a steel frame, within which is a movable horizontal beam called the cross-head, driven by a pair of helical threads that are powered by an electric motor (Fig. 3.2). The test piece is clamped between two sets of jaws. The lower set is rigidly attached to the bottom of the frame. The upper set, which has some lateral mobility, is connected, via the load cell, to the cross-head. The load cell houses a sensor that measures the load, in pounds, applied to the test piece. (Engineering machines, especially old or American ones, are often calibrated in English, rather than in metric, units.) When a test piece is clamped between the two jaws and the RAISE button is pressed, the helical threads start turning slowly. This action raises the cross-

Figure 3.3. Test pieces of a standard shape are cut from a sheet of material.

head, but too slowly for its movement to be detected by eye. As soon as the test piece starts being stretched, a load is registered by the pen on a chart recorder. The sensitivity can be adjusted so that the movement of the pen across the entire width of the chart paper represents 100-, 200-, 500-, or 1,000-pound loads. At the first setting, for example, a full-scale deflection of the pen represents a force of 100 pounds on the test piece. Stress at any particular point can be calculated by converting the pen reading into kilograms, multiplying by 9.81, and dividing by the original cross-sectional area of the test piece. The paper advances at a set rate, and so does the movement of the cross-head, so the two bear a constant relationship to one another. I have set the chart speed so that a 10-inch advance of the paper corresponds to an extension in the test piece of 1/2 inch. Other settings can be used, but I've found this one to be the most satisfactory for most purposes. You can calculate strain at any point by dividing the extension in the test piece by its original length. The resulting graph, penned on the chart paper, is therefore load, in pounds (y axis), plotted against extension, in inches (x axis). Although not a stress/strain graph, it is equivalent to one because load is directly proportional to stress, and extension is directly proportional to strain.

The test pieces are cut out of a sheet of material to a standard shape. They are narrow in the middle and flared at either end where they are gripped in the jaws (Fig. 3.3). As stress is inversely proportional to cross-sectional area, the stress is highest in the narrow neck region, which is where the break is designed to occur when the material is tested to failure. When test pieces break, they usually do so with a satisfying bang. Sometimes the break occurs at the shoulder region, usually because of a small nick incurred during the machining process. This kind of break, as we will see later, is a common problem when making test pieces from

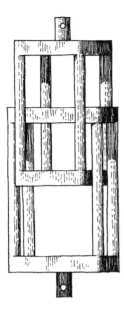

Figure 3.4. A cage device, used for compression testing in the Instron.

bone. Sometimes the break occurs in the gripped area, usually the result of over-tightening the jaws on a brittle material. In all of these cases the results are rejected because the data cannot be interpreted.

Before gripping the test piece in the machine, you need to measure the width and thickness of the neck region with calipers, so that you can calculate the cross-sectional area. Once the material is in the machine you can measure the distance between the edges of the jaws. This measurement gives the original length of the sample. The increase in length is read from the chart, hence strain can be deduced.

It could be argued that since most of the extension in the test piece occurs in the narrow region, where the stress is highest, the original length should be taken as the length of the narrow region, rather than the distance between the jaws. But what difference would this make to the results? The extension in the test piece would be the same, of course, regardless of which measurements were taken. But taking the larger measurement of the distance between the jaws as the original length increases the size of the denominator (strain = extension/original length), thereby underestimating strain. As we will see later, the Instron tends to overestimate strain, so this measurement procedure is advantageous.

By fitting a cage device between the jaws, you can set up the Instron for compression testing (Fig. 3.4). (A compression cell can also be used, which is more convenient, but it is not always available.) When the cross-head is raised, the floor of the cage moves toward the roof, so a cuboidal test piece placed inside the cage will be compressed. During tensile testing there is a clear-cut breaking point, and as soon as the test piece fails, the load upon it falls to zero. Compression testing, in contrast, often has no well-defined breaking point

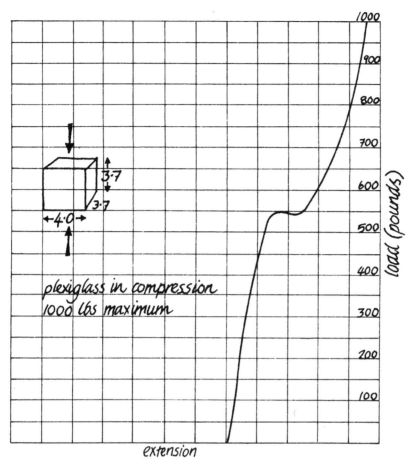

Figure 3.5. Instron output for a compressive test on a small cube of Plexiglas. For convenience, extension is depicted as increasing from left to right, the normal convention, rather than from right to left as actually happens in the machine. The same is true for subsequent charts.

because when the sample fails, the load, and therefore the compression, just goes on increasing. Failure may be marked by a sudden and rapid increase in strain. If the elastic behavior of the sample up to that point has been linear, or essentially linear, it may then become nonlinear. Both of these events (a rapid increase in strain and a departure from linearity) can be seen when compressing a small cube of Plexiglas to failure (Fig. 3.5). There is a large element of unpredictability in the behavior of the test piece during compression testing. The sample may buckle, crumble, split, flatten, or shear. Shearing, as noted in the previous chapter, involves a sideways movement due to lateral slippage of adjacent layers. Because of all this variation, compressive tests lack the repeatability of tensile ones, and the resulting values for Young's modulus, and for compressive strength, are far more variable than those for tensile tests. As a consequence, compressive data are less readily available in the literature. Therefore, if values are given for Young's modulus without specifying the loading regime, it is safe to assume the material was tested in tension rather than in compression.

Details of how to use an Instron machine are given in the appendix (p. 270). It is possible that you may not have access to such a testing machine. Even if you do, these labs can be quite time consuming if you are working in a large class. I will therefore depart from the usual format here. Instead of outlining a series of Instron experiments, I will reproduce some Instron chart outputs and work through their interpretation. I will start with the tensile testing of mild steel, because this test exemplifies most of the properties of materials discussed in the present and previous chapter.

Mild Steel Tested in Tension below the Breaking Point

If you look at the first part of the chart output (Fig. 3.6), you will see a simple diagram of the test piece, with the distance between the jaws, which is taken as the original length (51.7 mm), the cross-sectional dimensions of the neck (7.6 mm × 1.0 mm), and the full-scale deflection load (200 lb). It is important to record all of this information before the test commences. The objective of the first part of the experiment is to apply a small load that will not cause any permanent strain, and to use the results to deduce Young's modulus. The cross-head is lifted by pressing the RAISE button. The pen immediately sweeps across the chart paper, and it takes very swift eye-hand coordination, and some practice, to hit the STOP button before the pen plateaus out at the top of the chart. Pressing the STOP button halts the paper and maintains the same load on the test piece. The LOWER button is then pressed, bringing the cross-head down. The pen immediately falls as the load is removed from the sample, and it requires the same swift response to press the STOP button before the pen overshoots or undershoots the zero mark.

If you lay a ruler along the ascending arm of the graph, you will see that it is straight. This shows that the steel sample has a linear elasticity, as expected of a crystalline material. The descending arm is also straight. If you dropped a vertical line from the tip of the peak, you would find that it bisected the triangle formed by the ascending and descending limbs and the baseline of the chart. Measure the horizontal distance from the starting point of the pen to the midline of the triangle, and the corresponding distance to the finishing point of the pen. You will find they are the same (5.0 mm). The extension of the test piece during loading is therefore exactly the same as its contraction during unloading, showing that there was no permanent strain. Given that 10 inches along the horizontal axis is equivalent to 0.5 inches of true extension, and that the extension is represented by 5.0 mm along the horizontal axis, and given that the original length and cross-sectional dimensions are 51.7, 7.6, and 1.1 mm, you can calculate stress, strain, and therefore Young's modulus under the 200 lb loading regime.

You should have deduced that $E = 24.0 \times 10^9$ Pa, or 24.0 GPa. If your results are significantly different, let us work through the calculation together. Ten inches along the x axis of the chart is equivalent to 1/2 inch of extension. Con-

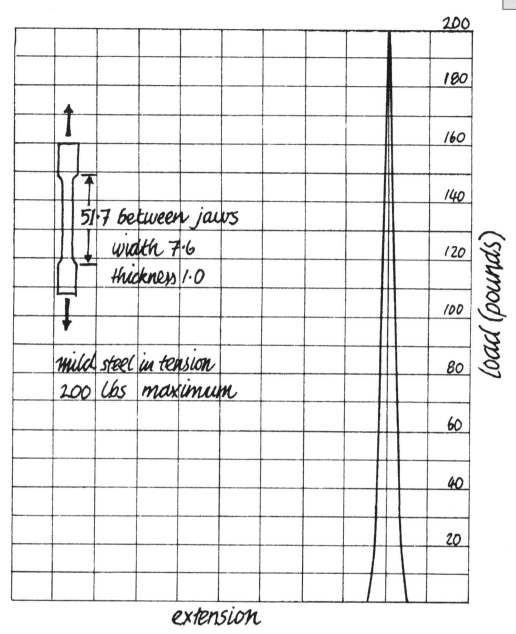

Figure 3.6. Instron output for tensile testing on a sample of steel. The load is not large enough to cause any permanent strain.

verting to metric units, 10 inches = 10.0×25.4 = 254 mm, and this represents 0.5×25.4 = 12.7 mm of extension. The actual distance along the x axis is only 5.0 mm, which is therefore equivalent to an extension of $5/254 \times 12.7$ mm. Strain = extension/original length = $5/254 \times 12.7/51.7$ = 0.004836. The force acting on the sample is $200 \times 0.45 \times 9.81$ N (1 lb = 0.45 kg). The cross-sectional area = 7.1×1.0 mm = 0.0076×0.001 m^2, therefore,

$$\text{stress} = \frac{200 \times 0.45 \times 9.81}{0.0076 \times 0.001}$$

$$\frac{\text{stress}}{\text{strain}} = \frac{200 \times 0.45 \times 9.81}{0.0076 \times 0.001} \times \frac{1}{0.004836}$$

$$= 24.02 \times 10^9 \text{ Pa, or } 24.0 \text{ GPa.}$$

Kaye and Laby (1995) give values closer to 200 GPa, which is an order of magnitude higher. We will return to the likely causes of the discrepancy later.

Mild Steel Tested in Tension, to Failure

The test piece used in this experiment was cut from an old filing cabinet and was therefore painted with gray enamel paint. This was fortuitous because the paint acted as a stress-coat, that is, a brittle material that is painted onto a test piece and that cracks at right angles to the direction of the *principal tensile strain,* the strain parallel to the direction of the load. Stress-coats are sprayed onto structures prior to testing to see where the points of maximum strain are located. The stiffness of the mild steel can be tested by applying finger and thumb pressure at one end. Mild steel does not feel particularly springy and requires a reasonable pressure to bend it.

The test piece, the same as used in the previous experiment, was loaded with a sufficient force to tear it apart. I found that a load of 500 lb was just large enough, but for subsequent testing on other samples I set the load to 1,000 lb, just to make sure. (There can be a fair amount of variation from one test piece to another, mostly because of differences in thickness in the sheets from which they are cut.) Now, looking at the Instron output (Fig. 3.7), identify the following features:

1. *Limit of proportionality (B).* Up to this point the relationship between load and extension (equivalent to stress and strain) is perfectly linear. Lay a ruler between A and B and check. This is the portion of the graph that would be used to calculate stress and strain, hence Young's modulus.

    Beyond the limit of proportionality there is a decrease in gradient, due to an increase in the extension, but it is not sudden.
2. *Yield point (C).* At this point there is a sudden increase in extension, hence in strain. Notice that there is an upper yield point (C) and a lower yield point (D), where the stress is slightly less than at the upper yield point. We need not be concerned with this, however.
3. *Elastic limit.* Somewhere between B and C is the elastic limit, the point up to which the material behaves elastically. From the elastic limit onward, the behavior is said to be plastic. The elastic limit is actually an ill-defined point and could be determined only by incrementally loading and unloading the sample to see at which point it just failed to return to its original length.

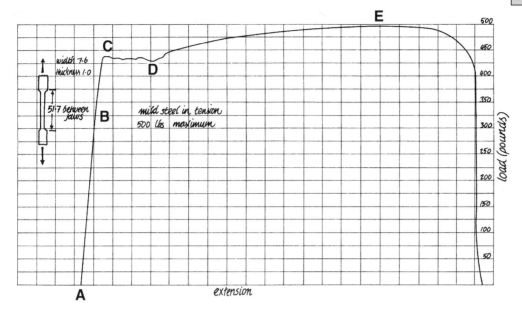

Figure 3.7. Instron output for a sample of steel, tested in tension to failure: (A) starting point; (B) limit of proportionality; (C) upper yield point; (D) lower yield point; and (E) breaking stress.

4. *Breaking stress (E)*. This is the maximum stress that can be tolerated by the material before it breaks. Notice that it is higher than the stress at the yield point. Stretching the metal actually strengthens it, and this process is called cold-working. Rolling and beating are the more usual ways of cold-working metals to strengthen them.

   Necking occurs just before the sample breaks, and it is about at this point that the paint cracks. Necking decreases the cross-sectional area, increasing the stress and causing the material to fail. Purists might argue that it is this *reduced* cross-sectional area that should be used in calculating the tensile strength. This calculation is sometimes done, and the result is termed

$$\text{actual breaking stress} = \frac{\text{breaking stress}}{\text{final cross-sectioned area}}$$

However, actual breaking stress is not used in engineering practice.

   Mild steel is obviously a ductile material because it has a well-marked phase of plastic behavior.

Calculate the tensile strength of the test sample. What value did you get? You should have calculated a value of 2.87×10^8 Pa (287 MPa), which compares quite well with the value given by Kaye and Laby (1995) ($4.3–4.9 \times 10^8$). If you got a different value, work through the following calculation: breaking force = $495 \times 0.45 \times 9.81$ N; cross-sectional area = 0.0076×0.001 m^2, therefore,

Figure 3.8. Examination of the broken ends of the steel test piece shows that they do not fit together exactly. This is because of the permanent strain; necking is also evident.

$$\text{tensile strength} = \frac{495 \times 0.45 \times 9.81}{0.0076 \times 0.001}$$

$$= 287{,}523{,}355.3$$

Now examine the broken test piece (Fig. 3.8):

1. If you tried joining the two pieces, you would find that they did not fit together very well. This is because of the permanent strain in the material. Necking would be obvious in both width and thickness.
2. You could test the stiffness of the metal in the stretched region using a gentle finger-and-thumb bending movement. If you compared this region with the unstretched region that was clamped in the jaws, you would find the latter was not quite as stiff. This is because cold-working has stiffened the steel – it has also increased its tensile strength.
3. Microscopic examination in the region of the break would reveal the cracks in the paint. Cracks in stress-coats run at right angles to the principal strain, which makes good sense when you think about it. You might therefore expect to see the cracks running at right angles to the direction in which the sample was pulled. However, the cracks run obliquely. They do so because the material shears when it breaks, and the shearing occurs at an angle of about 45°.
4. Microscopic examination of the shoulder region, where the sample was clamped, reveals a series of dimples. These dimples have been impressed into the steel by small projections on the jaws, designed for additional grip. The dimples farthest from the edge of the jaws are symmetrical, but they become elongated, in the direction of the pull, closer to the edge. What does this suggest?

Measurement Error

Calculation of Young's modulus gave a result that was about an order of magnitude lower than published values. However, the value obtained for tensile strength was correct. What does this result suggest as the source of measurement error? The evaluation of tensile stress involves measuring stress and the linear dimensions of the test piece, so we know that all of these variables were measured without significant error. Young's modulus involves measuring stress and strain, so the latter variable must be the source of error. Strain is evaluated from two measurements: the change in length and the original length. Since it is known that linear dimensions were measured accurately (calipers were used), the source of the error must be in the change in length.

Since the calculated value for Young's modulus is too low, the Instron must be overreading extension. One contributing source of error was evident from the elongated dimples in the shoulder region of the test piece. These dimples show that there was some slippage in the jaws. A second source of error is due to the elasticity of the machine itself, referred to as *machine compliance*. Although the Instron is constructed of heavy steel, none of these components are absolutely rigid. Consequently, as the sample is stretched, there will be a small amount of flexion in the machine, which will be read as part of the extension, though it will be minimal compared to that of the test piece. A third source of error, which is not apparent, is in the speed of the chart paper. This error can be verified by marking the chart and timing how long it takes to move 10 inches. It should take exactly one minute, the time it takes the cross-head to move 0.5 inches, but it actually takes slightly less time. This discrepancy therefore extends the y axis of the chart, exaggerating extension.

Being aware of some of the problems of testing simple materials like steel is good preparation for testing biological materials. Although the Instron machine I used gave inaccurate values for Young's modulus, the results obtained are useful for comparative purposes.

Tensile Testing of Annealed Brass to Failure

Brass is an alloy of copper and zinc. Annealing brass, or other metals, involves heating the material and letting it cool. This action changes its crystal structure, and hence, its mechanical properties. A brass test piece can be annealed simply by heating it up with a Bunsen burner and leaving it to cool. Annealing brass does not change its appearance, but it does make it far more ductile. If you had two test pieces, one annealed and the other not, you could tell them apart simply by trying to bend them. The untreated specimen would feel springy, like a saw blade, but the annealed one would bend very easily, demonstrating its plastic behavior. When annealed brass is loaded in tension in the Instron, it stretches by a surprisingly large amount. I have two test samples of annealed brass in front of me, one that has been pulled apart, and the other that is untested. They both started at the same length (166 mm), but the stretched one is now 192 mm long.

Examine the Instron output for annealed brass tested in tension to failure (Fig. 3.9). Notice the following features:

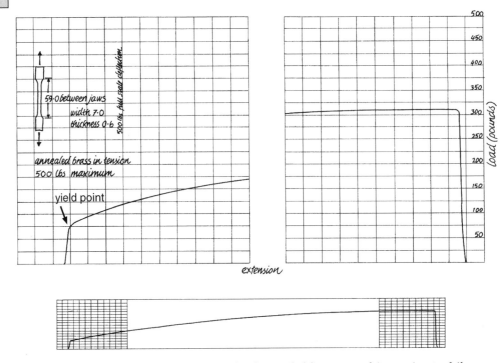

Figure 3.9. Instron output for a sample of annealed brass, tested in tension to failure. Because of the considerable extension, only the start and finish of the chart are shown in detail.

1. There is a linear portion, with a well-marked limit of proportionality. Beyond this point there is a sudden increase in extension. The limit of proportionality is therefore coincident with the yield point. The elastic limit is presumably also coincident with the limit of proportionality, but, as we have seen, we would have to determine this experimentally by incremental loading, up to the limit of proportionality.

2. Beyond the limit of proportionality is an extensive plastic phase, during which there is considerable deformation.

3. During the plastic phase a considerable amount of hardening occurs (cold-working). Notice that the breaking stress occurs at a loading that is several times higher than the loading at which the material yields. Yielding occurred at a load of 80 lb, whereas the sample did not fail until the load reached 310 lb.

The linear portion of the load/extension graph is rather too small to be able to obtain an estimate of the extension for calculating Young's modulus. However, you will be able to deduce the tensile strength, given the cross-sectional dimensions of the sample and the clearly marked breaking point at 310 lb.

If you examined the broken test pieces, you would find the following:

1. The surface looks, and feels, textured in the stretched region. This is because of the elongation and displacement of the crystals of material (metallurgists refer to crystals as grains).

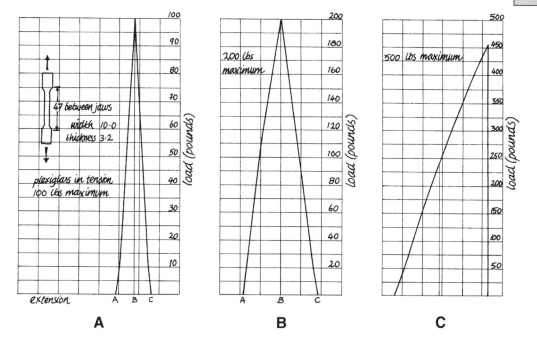

Figure 3.10. Instron output for a sample of Plexiglas, tested in tension, with increasing loads. Initially, the material behaves elastically (A). As the load increases, the elastic behavior becomes nonlinear (B). Testing to failure shows markedly nonlinear elasticity (C).

2. The cold-worked region feels stiff compared with the portion that was gripped in the jaws.
3. The two broken ends fit together moderately well. Necking is not so obvious as it was in the mild steel, but it can be seen in edge view, especially under the microscope.

Annealed brass is an extremely ductile material.

Plexiglas Tested in Tension

The Instron output shows three consecutive tests on the same sample of Plex-iglas, with the loading set at 100 lb, 200 lb, and then 500 lb (Fig. 3.10). In the first test the loading has not exceeded the limit of proportionality and the load/extension graph (hence the stress/strain graph) is linear. The material behaves elastically within this loading range, which can be confirmed by bisecting the triangle and checking its symmetry. Thus the increase in length during loading (*AB* in Fig. 3.10A) is the same as the decrease in length during unloading (*BC* in Fig. 3.10A). Because of space limitations these, and subsequent Instron outputs, are less than full size. This must be accounted for when taking measurements from the illustrations. The full height of the Instron chart is 6 inches, whereas in Figure 3.10 it is only 3 inches. Thus the chart extension in Figure 3.10A, as measured from the illustration, is 4.7 mm (AB), so the actual chart extension is $4.7 \times 6/3 = 9.4$ mm. (Remember that 10 inches of chart extension represents 0.5 inches of true extension.) Deduce Young's modulus of the sample.

During the second test, the first portion of the extension exemplifies linear elasticity, but the graph becomes nonlinear (Fig. 3.10B). If you lay a ruler along the first part of the load/extension graph you will see that it is linear up to a load of approximately 115 lb. The triangle appears to be symmetrical, showing that there was no permanent strain and that the material is behaving elastically. The departure from nonlinearity is best seen in the third test, where the load/extension graph is markedly curvilinear (Fig. 3.10C). The material failed at a loading of 460 lb. Notice that there is no yield point, so there does not appear to be any plastic deformation. However, this conclusion could be verified only by incremental loading and unloading, to see whether the sample returned to its original length after each trial. Deduce the tensile strength.

If you examined the broken test piece, you would see that there was no evidence of necking, which would be confirmed by a microscopic examination. Since there does not appear to be any evidence of plastic deformation, the material may be tentatively described as nonductile, that is, brittle.

Evaluation of *E* and Tensile Strength of Plexiglas in Tension

From the load/extension graph of the first test (Fig. 3.10A) you would have been able to measure a paper extension of 7 mm (*AB*), produced by a loading of 93 lb. The original length of the test piece was 47 mm, and its cross-sectional dimensions were 10 mm by 3.2 mm. From these data you should have obtained a value of 1.72×10^9 Pa, or 1.72 GPa for Young's modulus. Published values give 6.16 GPa, which is higher, but that is because the Instron overestimates extension. Your value for the tensile strength should have been 0.63×10^8 Pa, or 63.46 MPa; the published values for plastics range from 28 to 70 MPa.

Plexiglas Tested in Compression

Plexiglas is quite strong in compression, stronger than it is in tension, so large stresses are required to cause failure. Consequently, fairly small cubes have to be used during compression testing, otherwise they cannot be made to fail. (Strictly, they should be referred to as cuboids because their length, width, and height measurements are not exactly the same, but I will not do this.) I used cubes that were only about 4 mm across. Although satisfactory for investigating failure, these small cubes give large errors when evaluating Young's modulus. I therefore used a larger cube for evaluating *E,* loading this cube to 1,000 lb (Fig. 3.11). If you check with a ruler you will see that the elasticity of the sample is almost perfectly linear, and the symmetry of the loading and unloading graphs shows that there is no permanent strain, so the material does not undergo plastic deformation. Given that the height of the cube was 17.2 mm and its other dimensions were 20.8 and 10.0 mm, evaluate Young's modulus.

The next Instron output (Fig. 3.12) shows the results of four experiments,

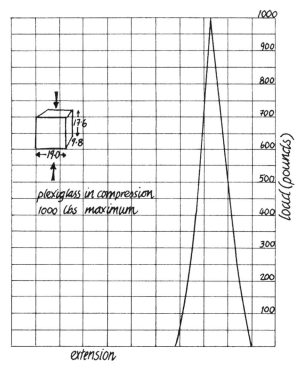

Figure 3.11. Instron output for a sample of Plexiglas, tested in compression. The elastic behavior is almost perfectly linear.

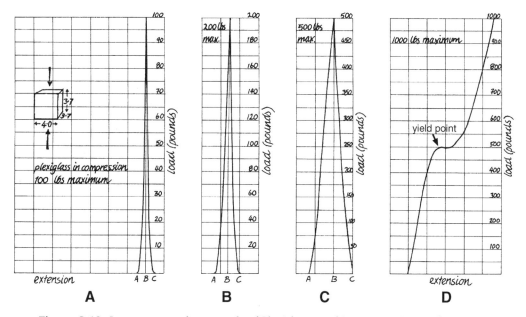

Figure 3.12. Instron output for a sample of Plexiglas, tested in compression, with increasing loads. Initially, the material behaves elastically, which is linear (A, B). With increasing loads the elasticity becomes slightly nonlinear, and there is some permanent strain (C). There is a well-marked yield point when the material is loaded to failure (D).

51

all on the same cube. In the first two records the cube was loaded with 100 lb and 200 lb, respectively. Use of a ruler will confirm that elasticity is linear, and the symmetry of the triangles formed by the loading and unloading graphs shows that there was no plastic deformation.

The third record shows a load of 500 lb. The loading graph is slightly curvilinear, and the triangle is obviously asymmetrical ($AB > BC$), showing that there has been some permanent strain. The 500 lb load therefore exceeded the elastic limit, and the cube is not quite as tall as it was before the test, having undergone some slight plastic deformation. The elastic limit is probably somewhere between 200 and 300 lb.

The maximum load was increased to 1,000 lb for the last experiment, and the test was continued until failure. Notice that there is a well-marked yield point, indicating a sudden increase in extension (thus strain). It was at this point, at a load of 550 lb, that the cube sheared. This shearing was accompanied by a rise in load (thus stress) that would have continued as the distorted cube became further compressed. It was therefore necessary to turn off the machine and lower the cross-head. There is, of course, no well-defined breaking point as there was when testing in tension, where failure was marked by the yield point. There is no cold-working as there is in metals. Deduce the compressive strength of Plexiglas from these results (Remember to allow for the reduction in the size of the illustration, as explained on page 49.).

What value did you get for your estimate of Young's modulus? You should have obtained something in the order of 0.53×10^9 Pa, which is lower than that obtained in tension (1.72×10^9). This indicates that Plexiglas is somewhat stiffer in tension than in compression, though the compliance of the machine is likely to differ between the two loading regimes, so we should not read too much into the results. As taking measurements from Instron charts is not very accurate, you should not be concerned if your results are different from those given. Provided you are within about 50 percent of the given value, you are probably close enough.

Your value for the compressive strength should have been in the order of 1.77×10^8 Pa, or 177 MPa, which is higher than the tensile strength value of 0.63×10^8 Pa, or 63 MPa. Plexiglas therefore appears to be somewhat stronger in compression than in tension.

If a Plexiglas cube were rotated through 90° and its mechanical properties reassessed, there would be no appreciable difference in its elastic modulus (the same would also be true of its strength). Like many other materials, including metals and ceramics, Plexiglas is isotropic. But such is not the case with wood or bone. Since these materials are anisotropic, you must be careful when loading them into test machines to make sure of their correct orientation.

Wood Tested in Tension and in Compression

Aside from being anisotropic, wood behaves much like Plexiglas when loaded in tension and compression. Thus, when loaded with forces that do

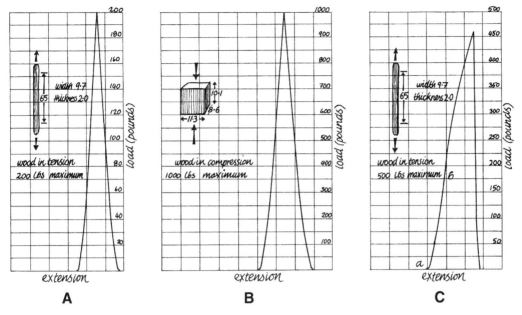

Figure 3.13. Instron output for a sample of wood, tested in tension (A, C) and in compression (B). Initially, the elasticity is linear (A, B), but when the load increases, the elastic behavior becomes nonlinear (C). Plastic deformation occurs sometime after the limit of proportionality b is exceeded.

not exceed its elastic limit, wood exhibits linear elasticity (Fig. 3.13A, B). When tested in tension to failure, the sample's elasticity is initially linear (a–b in Fig. 3.13C), but it then becomes nonlinear. Presumably plastic deformation occurs somewhere after the limit of proportionality (b in Fig. 3.13C), but this could only be established by incremental loading and unloading and checking each time to see if there had been any permanent strain.

The anisotropic properties of wood can be graphically shown by changing the orientation of the test piece. A small cube of wood was first tested in compression, with the grain running parallel to the direction of the force (Fig. 3.14A). The Instron was set to give a full-scale deflection of 1,000 lb, which was not enough to cause the cube to fail in this orientation. The cube behaved with a linear elasticity, returning to its original length when the load was removed (Fig. 3.14A). The sample was then turned through 90° and loaded again. Initially the elasticity of the sample was linear, but that ended at a load of about 450 lb, and the sample yielded at a load of about 540 lb, undergoing plastic deformation. If a vertical line is dropped from the limit of proportionality, you can get an idea of the amount of extension that was recorded during the linear elastic phase – it is represented in the figure by the distance c–d (Fig. 3.14B). This extension is much greater than the corresponding extension when the specimen was loaded parallel with the grain (Fig. 3.14A, a–b). This experiment clearly shows that the wood undergoes greater extension, and therefore has a larger strain, when loaded perpendicular to the

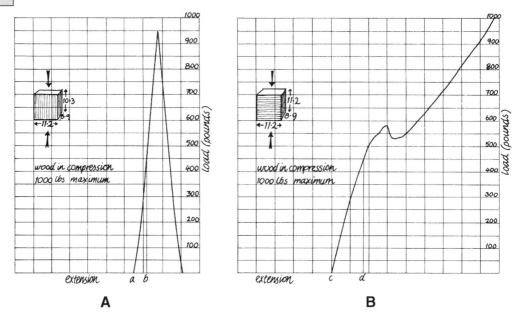

Figure 3.14. Instron output for a sample of wood tested in compression. When loaded parallel to the grain (A), the wood is stiffer and stronger than when loaded perpendicular to the grain (B).

grain rather than parallel to it. Consequently, the Young's modulus is lower (remember that E = stress/strain). The experiment also shows that wood is much weaker in compression when loaded against the grain than with the grain. The same is also true for wood loaded in tension – it is stiffer and stronger loaded parallel to the grain than against the grain.

The Instron values obtained for wood (loaded parallel with the grain) for Young's modulus in tension and in compression are 6.4 GPa and 0.6 GPa (64.0 and 5.9×10^8 Pa), respectively (though, again, these results are likely to be affected by differences in machine compliance). Although these values are inherently inaccurate, they are still useful for comparative purposes. The tensile and compressive strengths (again loaded parallel to the grain) are 105 and 44 MPa, respectively. Wood is therefore both stiffer and stronger in tension than in compression, and it is always stiffer and stronger loaded with the grain than against it.

Compared with wood, steel is much stiffer (Instron value of E for steel = 24 GPa) and has a higher tensile strength (Instron value for tensile strength of steel = 290 MPa). But steel is also considerably more dense (density = 7,800 kg/m³, compared with 750 for wood). Therefore, on a weight-for-weight basis, wood compares very favorably with steel, both in stiffness and tensile strength. This comparison can be expressed by the terms *specific modulus* and *specific strength,* where specific modulus = Young's modulus/density, and specific strength = tensile strength/density. As we know, both Young's modulus and tensile (and compressive) strength are measured in pascals, but if these are divided by density, the units change. You can avoid this unit change by dividing by *relative density*, which is the density of the material compared with that of water (this used to be referred to as specific gravity). Wood has a relative density of 0.7, while that of steel is about 7.8. I will refer to the result of dividing Young's modulus and tensile strength by relative density informally as "relative modulus" and "relative strength," to avoid confusion with the formal terms of specific modulus and specific strength (Gordon 1978 uses the term "specific Young's modulus" for what I refer to as relative modulus). If we use the values for Young's modulus and tensile strength given in Kaye and Laby (1995), the relative modulus of wood and steel are very similar (about 19 GPa and 27 GPa, respectively). However, the relative strength of wood is much higher (86–157 MPa) than that of steel (55–63 MPa). If we wanted to build an aircraft we would obviously use wood in preference to steel. Wood was used in the construction of early aircraft, but this practice has now been abandoned in favor of lightweight alloys and plastics. The advantages of these materials over wood include their durability (they do not rot) and the convenience of fabrication. Wood was used exclusively for shipbuilding during the early days of sail, but steel is now used in preference, except for the smallest vessels. This is because steel construction is more practical, large timbers being all but impossible to obtain anymore. Wood also has the disadvantage of swelling when wet, and also of rotting.

Bone Tested in Tension and in Compression

One of the biggest problems in conducting mechanical tests on bone is preparing the samples. There is also the problem of clamping the tensile test pieces tightly enough to prevent slipping without cracking them. Tensile test pieces for my class were machined to the same shape as the metal ones but were made much smaller (about 60 mm long compared with 180 mm). All of the samples, which were of compact bone, were cut with their long axes parallel to the long axis of the bone (the cannon bone of a horse was used). Small cubes of edge about 4 mm were machined for compressive testing. Because bone is anisotropic, care was taken to maintain the original orientation of the sample during machining, and the top and bottom of each cube was marked with an indelible dot.

When tested in tension with loads less than that required to break them, the bone samples exhibited linear elasticity (Fig. 3.15A). The symmetry of the loading and unloading graphs showed there was no plastic deformation. Nor was there any indication of a plastic phase when loaded to failure; the bone behaved as a typically brittle material (Fig. 3.15B). Similar results were obtained by earlier researchers, and it was generally concluded that bone did not exhibit plastic behavior. However, Burstein, Currey, Frankel, and Reilly (1972) showed that this conclusion was not true, and that bone underwent considerable plastic deformation prior to breaking (Fig. 3.16). The reason why this plastic deformation had been missed before was attributed to the use of dry bone samples. Burstein and his colleagues obtained fresh bone and stored it in a freezer until it was used. By keeping it wet with a dilute saline solution

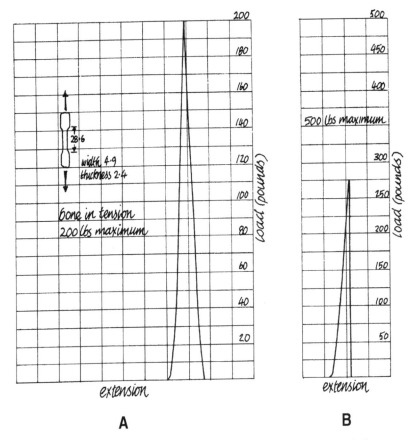

28·6

width 4·9
thickness 2·4

bone in tension
200 lbs maximum

500 lbs maximum

load (pounds)

load (pounds)

extension

extension

A

B

Figure 3.15. Instron output for a sample of horse cannon bone, loaded in tension. With loads that are insufficient to cause failure, the elasticity is linear (A). When loaded to failure, there is no indication of a plastic phase; the behavior is typical of a brittle material (B).

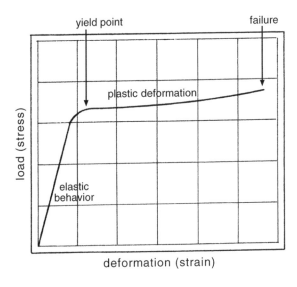

yield point

failure

plastic deformation

load (stress)

elastic behavior

deformation (strain)

Figure 3.16. Fresh bone exhibits considerable plastic deformation when loaded to failure, as shown by Burstein, Currey, Frankel, and Reilly (1972). The sample was loaded in tension.

during their experiments, the bone was kept as close to living conditions as possible.

Another major difference between their experimental design and the one used here was that they attached an instrument to the bone for measuring strain directly, thereby eliminating the machine-borne errors of the Instron. Although the behavior of the bone was ductile prior to failure, there was no evidence of necking (Burstein, Reilly, and Frankel 1972). After yielding, the bone undergoes a considerable increase in strain, and some increase in stress, too. However, as Currey (1984) points out, the stress at the point where the bone fails in tension is only marginally higher than at the yield point, so its plastic behavior makes little difference to the end result. If the plastic behavior of bone were like that of annealed brass, where the stress at the breaking point is several times higher than at the yielding point, it could make a significant difference to the outcome. Thus an incident that exposed a bone to stresses large enough to cause it to yield would not necessarily cause it to break. If a bone is stressed to the point of entering into the plastic phase, but without actually breaking, some damage is probably sustained in the form of microfractures, which have to be repaired. We will discuss these microfractures later, in Chapter 5.

The Instron value that we obtained for the tensile strength of bone was 103 MPa, which is not far from published values (Reilly and Burstein 1975 give an overall value of 133 MPa for human bone). However, our values for Young's modulus were predictably underestimated. Thus the Instron value for *E* in tension was 5.6 GPa, compared with average values of 17 GPa for human femur and 24 GPa for bovine femur (Reilly, Burstein, and Frankel 1974). As the bone samples used in our classes were all cut parallel to the longitudinal axis of the bone, that is, with the grain, the tensile testing gave no opportunity to investigate the anisotropic properties of the bone. However, we performed the compressive tests in both directions.

We oriented a small bone cube ($3.4 \times 3.6 \times 2.1$ mm) with its grain parallel to the compressive stress (that is, with the grain) loaded to 200 lb, and then unloaded (Fig. 3.17A). The sample behaved elastically, as shown by the symmetry of the loading and unloading graphs, and was also linear. We repeated the test with a load of 500 lb, with similar results. When we set the Instron to a maximum load of 1,000 lb and repeated the test yet again, the sample yielded at a load of 690 lb (Fig. 3.17C). If you lay a ruler against the elastic portion of the chart, you will see that it is linear, up to a loading of about 600 lb, which marks the approximate position of the limit of proportionality. We set up a second, and somewhat larger, bone cube ($5.0 \times 5.0 \times 4.7$ mm) against the grain and loaded and unloaded, first at 200 lb, then at 500 lb (Fig. 3.17D, E). In both cases there was a marked asymmetry in the loading and unloading graphs, showing that there had been some plastic deformation and therefore some permanent strain. When the load was increased to 1,000 lb, the cube did not fail, but there was marked asymmetry again, showing that there had been some plastic deformation. We then turned the cube through 90° so as to load it with

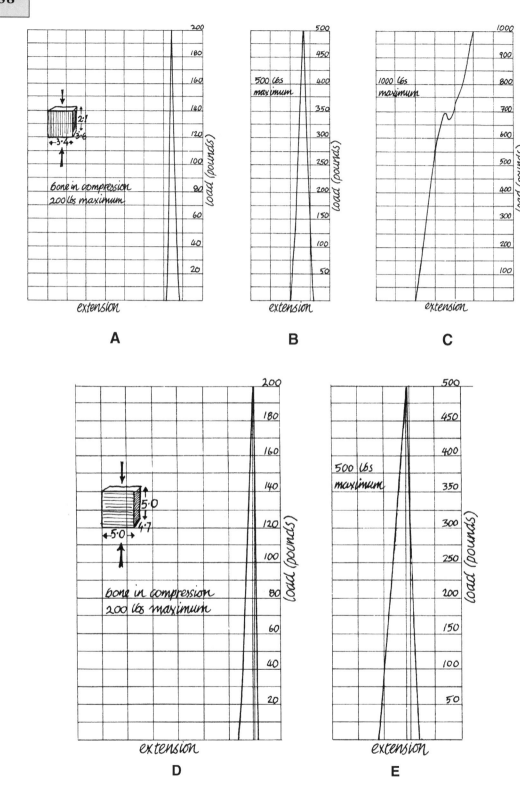

A

B

C

D

E

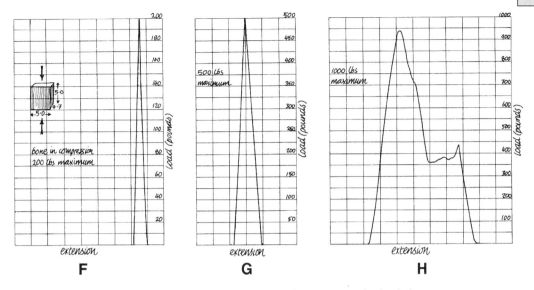

Figure 3.17. *(Opposite and above)* Instron output for bone samples loaded in compression, parallel to the grain (A–C; F–H) and perpendicular to the grain (D–E).

the grain, and we repeated the loading regimes. This time the behavior of the bone was linearly elastic at loads of 200 and 500 lb (Fig. 3.17F, G), but it failed at a load of 935 lb (Fig. 3.17H). Compression tests on other cubes showed that bone was somewhat stronger when loaded with the grain than against it, so its failure at a load below that which it had survived when loaded against the grain was unexpected. Presumably the reason for this failure was that the previous testing, in exposing the bone to plastic deformation, had weakened it.

Reilly, Burstein, and Frankel (1974) found no significant differences between the Young's modulus when measured in tension or compression. However, bone is stronger in compression than in tension (193 MPa and 133 MPa respectively; data from Reilly and Burstein 1975), and it is also stronger when loaded with the grain than against it, both in tension and in compression. Thus, when bone is loaded against the grain, tensile strength falls from 133 MPa to 51 MPa, and compressive strength falls from 193 MPa to 133 MPa (data from Reilly and Burstein 1975).

Although the Instron tests did show some plastic deformation during compression, it was not extensive, largely because the samples were dry. Investigations on wet samples have shown that bone undergoes an extensive plastic phase during compression, just as it does during tension (Fig. 3.18). During tensile testing the stress increases after the yield point is reached (Fig. 3.16), but it decreases during compressive testing.

We have already seen how the state of hydration affects the mechanical properties of bone. I now want to briefly discuss strain rate and how our laboratory results differ in this respect from what might be expected in life. Instruments like the Instron apply a strain to the test piece rather slowly. For example,

Figure 3.18. Fresh bone exhibits considerable plastic deformation when loaded in compression, as it does when loaded in tension (Fig. 3.16). However, in contrast, the stress decreases in compression after the yield point is reached (after Reilly, Burstein, and Frankel 1974).

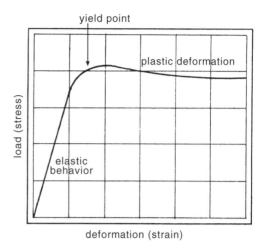

during one of the tensile tests to failure, it took 2.6 seconds to apply a breaking load. In life, an animal would load a bone in a fraction of a second, so the Instron loading regime is far removed from reality. However, there are hydraulically driven test machines that can load and unload samples at high strain rates. It has been found that tensile and compressive strengths, and Young's modulus, increase with increasing strain rates, as reported in Currey (1984). Laboratory tests at slow strain rates therefore underestimate bone strengths and elastic moduli.

Fatigue Fracture

All of the testing discussed so far involves incrementally loading a material, either in tension or in compression, until it breaks. The maximum stress in the material at the point of failure is thereby defined as its tensile or compressive strength. However, if the material is repeatedly loaded and unloaded, failures can also take place at stresses that are well below these levels. This phenomenon, where repeated loadings at levels below breaking stress can cause failure, is referred to as *fatigue fracture.*

Many of us use fatigue fracturing to break things when an appropriate tool is not close at hand. When a new credit card arrives in the mail the old one is more often broken by repeated bending than by dispatching it with scissors. Stubborn sardine can lids are often dealt with in the same way. You can carry out a simple fatigue fracture experiment with a paper clip. If you open out a paper clip by separating the two loops, you apply a torsional stress to one of the bends. The stress is below the breaking stress of the metal, so it does not fail, but if you repeatedly bend it back and forth, it breaks. Fatigue fracture is probably a common cause of failure, both in animate and inanimate structures, and was responsible for a series of aviation disasters that began in England almost half a century ago.

Life in England during the early fifties was not a lot of fun. Food and other essentials were still on ration, and the economy was struggling to pull itself out of the rubble of the recently ended war in Europe. Billboards and newspaper advertisements encouraged people to "Buy British," and national pride, if not national wealth, was at a high. One of the brightest jewels in Britannia's crown was the de Havilland Comet, the world's first jet airliner. The Comet entered into scheduled service with the British Overseas Air Corporation on May 2, 1952, with an inaugural flight on the London-Johannesburg route. All went splendidly well at first, but within a year two Comets had crashed during takeoff, the first without loss of aircraft or passengers. The problem, once identified as a wing-stalling condition, was soon rectified. Then, on May 2, 1953, a third Comet crashed as it reached its cruising altitude of 35,000 feet, after taking off from Calcutta. This time the aircraft disintegrated in the air, raining debris over the land for a distance of eight miles. The loss was attributed to a thunderstorm, but then two more Comets crashed under similar circumstances. The aviation authority ordered the grounding of the entire Comet fleet.

Both incidents occurred over the Mediterranean, shortly after taking off from Rome. The fact that the aircraft had disintegrated in flight suggested a bomb, but bodies recovered from the sea showed no residues of explosives, though there was evidence of an explosive decompression of the cabin. The decompression raised the possibility that an engine had disintegrated, puncturing the pressurized fuselage with high-speed turbine fragments, thereby causing it to burst like a balloon. One of the crashed Comets was salvaged from the bottom of the Mediterranean by the Royal Navy. Remarkably, 70 percent of the wreckage was recovered. The fragments were painstakingly reassembled in an aircraft hangar in England. The clues provided by these fragments turned attention away from engine failure and toward the possibility of metal fatigue.

Aircraft had been flying with pressurized cabins since 1940, but they were piston-engined aircraft and operated at much lower altitudes than the Comet's 35,000 feet. The larger pressure differences between the outside and the inside of the cabin at these higher altitudes placed larger stresses on the skin. Furthermore, each cycle of pressure change during ascent and descent caused flexing of skin – like a balloon being blown up and partially collapsed again. The Comet had a range of only 1,750 miles, requiring frequent refueling stops and therefore frequent repressurizing cycles. To simulate this situation, the crash investigators acquired a Comet, built a tank around the fuselage, filled both the fuselage and the tank with water (to contain any possible explosion), and repeatedly pressurized and depressurized the cabin. After just over 3,000 cycles, approximately equivalent to almost two years of airline service, the fuselage failed. Comparisons between the test fuselage and the remains of the salvaged aircraft revealed striking similarities confirming that metal fatigue had been the cause of the crash. The problem was rectified and the Comet reentered service in 1958, but by then the Boeing 707 was already operational. British aviation never recovered its lead. (See Demster 1958 for a good account of the Comet and its troubles.)

Fatigue fractures are probably fairly common in terrestrial vertebrates, especially in the feet and lower limb elements, and many of us have had personal experience with them (see Alexander 1984a for a useful discussion). Sometimes they remain as small, frequently painful cracks, often passing undetected during x-ray examination. (Doctors refer to them, inappropriately, as "stress fractures," as if other fractures were not caused by stress!) At other times fatigue fractures cause the entire bone to break, often catastrophically. This may occur when an antelope is fleeing from a lion, but it can also occur when animals are not maximally loading their bones. Fatigue fractures commonly occur in racehorses, which is not surprising, and there are anecdotal accounts of bones breaking during races with a loud report like gunfire.

We have seen that tensile and compressive strengths increase with strain rate, so too does the fatigue life of a bone, which is the time taken for a bone to break during fatigue testing. For example, increasing the strain rate from 30 Hz (cycles per second) to 125 Hz increases the number of cycles it takes to break a bone by two or three times (reported in Currey 1984).

Tendons in Tension

Tendons, the tough cords that connect muscles to bones, were the most difficult for our class to test because of the problem of trying to secure them in the jaws of the Instron, partly due to the fact that they are so slippery. Tendon was also the most interesting material our class tested, because of an unusual mechanical property described as relaxation. Since tendons are flexible and are loaded in tension, compressive testing is not relevant.

Some of the tendons we tested were from the leg of a horse and were about as thick as a finger. Others, from the leg of a calf, were much smaller, about 4 mm wide and 2.2 mm thick. After gripping the tendon in the Instron, we set the load to the 100 lb range. The sample was initially stressed to a loading of about 50 lb, then the machine was reversed, dropping the load to zero (Fig. 3.19A). The extension was markedly curvilinear, but the behavior was perfectly elastic. The same result was obtained at a loading of 60 lb (Fig. 3.19B). We repeated the operation but this time pressed the STOP button when the load reached about 70 lb, thereby halting the chart paper and holding the sample at the same extension and therefore the same strain. Although the strain was kept constant for the next minute, the load, hence the stress, steadily declined (Fig. 3.19C). This phenomenon, where the stress in a material declines while the strain remains constant, is termed *relaxation*. Relaxation results from a rearrangement of the molecules. A similar but converse phenomenon is *creep*, where the stress remains constant but the strain changes. Wood is a material that exemplifies creep, and I have a classic example of the phenomenon in my living room. My house, like so many other modern ones, was not constructed with the best materials and the floors are made of ply-wood, rather than of thick planks. We have two china cabinets, both filled with heavy crockery and glassware, and over the years the floor has sagged under the constant heavy load. The stress has remained essentially the same, but the strain has increased, due to creep in the wooden floor. A somewhat similar thing happened in the driveway of my old house. When I rented the house during a sabbatical trip abroad, the new tenant used to leave his car parked on the asphalt driveway. Asphalt may seem like a perfectly good solid, but in reality it is a very viscous liquid, and if a sufficient stress is applied for a length of time, in this case the weight of a car, the strain increases. As a result, the car slowly sank into the driveway, leaving deep ruts. The phenomenon can be demonstrated with a barrel of pitch and a brick, or even a heavy weight and a pitch-lined dissecting dish. In both cases the object slowly sinks into the pitch. Eventually the brick would sink to the bottom of the barrel. Because materials like tendon and wood behave somewhat like viscous mate-rials, they are described as being *viscoelastic,* and relaxation and creep are examples of viscoelasticity.

In the last tendon experiment, we started the Instron, still set at 100 lb, and loaded the sample until it failed. This time the load reached a level of almost 60 lb, and then the sample began to yield (Fig. 3.19D). Yielding was a protracted phase in which the cross-head continued to rise, stretching the tendon to greater lengths, but the load steadily fell. As the tendon became stretched even further, individual fibers became apparent at the surface, and these started breaking. The stretching also caused a decrease in the thickness of the tendon, resulting in its slipping in the jaws, so the test did not con-tinue to the complete failure of the material.

Estimates of Young's modulus can be obtained by drawing a line through the straightest portion of the load/extension graph. The results we obtained were about 0.4 GPa. Bennett, Ker, Dimery, and Alexander (1986) obtained estimates for various mammalian tendons by drawing tangents at various points on the curvilinear load/extension graphs. The values ranged from about 1 to 2 GPa.

We were able to obtain some indication of the tensile strength by consid-ering the largest load the tendon experienced; our results gave values of 44 MPa. Bennett et al. (1986) also had difficulties measuring tensile strength because so many of the specimens failed without a clean break. They con-cluded that the tensile strength of tendon was at least 100 MPa.

Rubber in Tension and Compression

The testing of rubber in tension in the Instron is not very satisfactory because of its inordinately large strain. Even when we clamped a short test piece in the jaws, the extension was so large, and the load so small, that

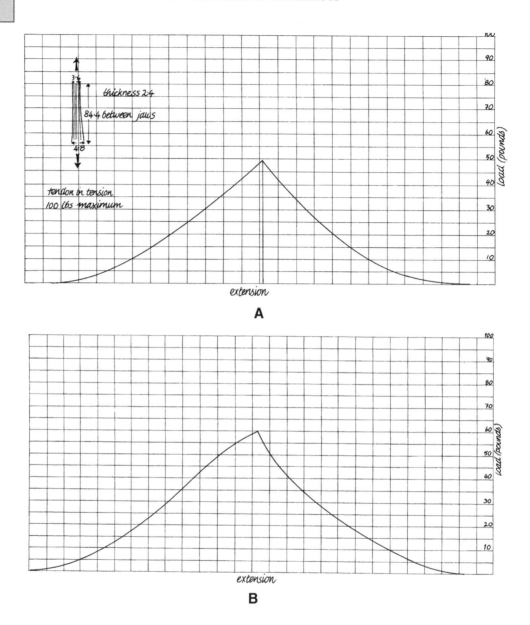

reams of chart paper were used when the Instron was at its normal setting. The resulting chart remained parallel with the base line for much of its length, and it took a long time to reach the breaking stress. Consequently, it was impractical to estimate Young's modulus from the results; published values are in the order of 7 MPa. However, the tensile strength can be determined, and we obtained values of 0.40 MPa. The time taken to reach the breaking stress, and the amount of chart paper used, are reduced by increasing the speed of the cross-head and decreasing the chart speed.

Compression testing, even with all its inherent problems, was far more

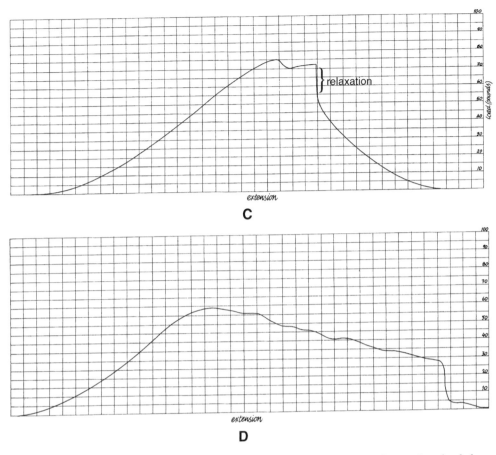

Figure 3.19. *(Opposite and above)* Instron output for a sample of horse leg tendon, loaded in tension below critical levels (A–C) and to failure (D). The elasticity is nonlinear (A, B). When the sample is loaded and the strain is kept constant by halting the machine, the stress slowly decreases, a phenomenon termed relaxation (C). Yielding was protracted, and the test did not continue to failure because of slippage in the jaws of the Instron (D).

satisfactory, and we found that a convenient material was a cube cut from an eraser. Erasers are not made from pure rubber but have various additives including an abrasive and a pigment. Consequently the material is somewhat brittle, which explains why erasers can often be broken by bending them in half. The Instron was first set to 100 lb, and the sample loaded and unloaded (Fig. 3.20A). The behavior was elastic but markedly curvilinear, so it is not very meaningful to evaluate its Young's modulus, unless for a specific stress range. When loaded to failure, the samples usually cracked, giving a well-defined break (Fig. 3.20B). Estimates for the compressive strength were 33.4 MPa.

If you have access to a suitable testing machine, you will be able to repeat these experiments yourself. Otherwise, you have probably acquired a sufficient

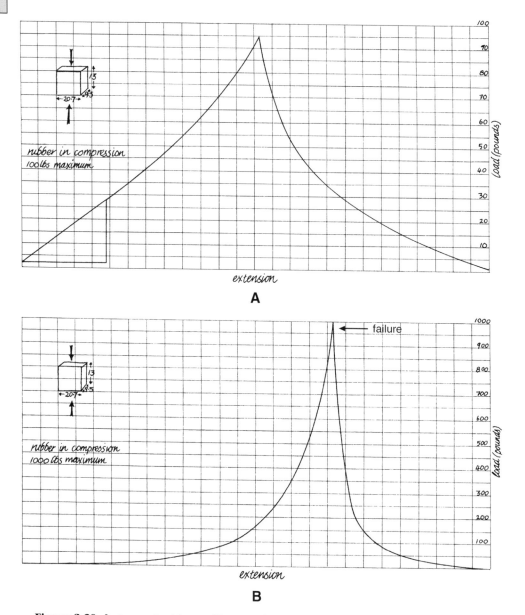

Figure 3.20. Instron output for a rubber sample (eraser), loaded in compression below critical levels (A), and to failure (B).

background in the nuances of testing materials and have a good enough idea of tensile and compressive strengths to move on. The subject of the next chapter is the mechanics of how things break.

4 How Things Break

I f you looked up the values for the tensile strengths of nylon, glass, cast iron, rope, aluminium, certain plastics, wood, and copper, you would find that they were all about the same. But we know from experience that some of these materials are much easier to break than others. Wine glasses break easily enough while being washed up, let alone while being dropped onto the floor, but you would be hard-pressed to break a plastic beaker, however careless you were. Tensile strength obviously tells us little about the resistance to breaking. So what does tensile strength tell us?

Just think how tensile strength is measured. A load is applied to a piece of material, and this load is increased until the material breaks. It is a static loading situation, in which a stationary force is applied to the cross-sectional area of the material. The units of tensile strength, as we have seen, are newtons per square meter, or pascals. It is true that things do break during static loading. However, if you tried to think of examples of things breaking in everyday life, you would find that most of them involved dynamic loading, where things are subjected to moving forces, that is, to work. To take a simple example, imagine a brick resting on a sheet of glass. If the brick had a mass of 2 kg, the static force acting on the sheet of glass would be 2×9.81 N. The glass does not break. Suppose the brick is now lifted 1 m above the glass and dropped. Work is force multiplied by distance, so the work done on the glass by the brick is $1 \times 2 \times 9.81$ joules (J). The glass breaks. The work required to break the glass is described as the work of fracture. The *work of fracture* is the amount of work that has to be performed upon a square meter of material to break it. The units of work of fracture are therefore J/m^2. It would be impossible to evaluate the work of fracture by dropping a brick onto a sheet of glass. This is because the glass would break in such a complex way that it would be impossible to determine the exact surface area over which the work was performed in breaking it. However, suppose it were possible to suspend the sheet of glass from one end so that it hung vertically. Suppose also that a steel plate were attached along the entire length of its free end such that when a brick was dropped onto this plate the work would be transferred evenly to the cross-sectional area of the glass sheet. It would now be possible to evaluate the work of fracture by dropping the brick onto the steel plate, from increasing heights, until the glass broke. Suppose the sheet of glass were half a meter wide and 6 mm thick, and that the 2 kg brick just broke the glass when it was dropped from a height of

97 mm. The work done by the falling brick would be: $2 \times 9.81 \times 0.097$ J $=$ 1.903 J. This works on a cross-sectional area of 0.5×0.006 m^2 $= 0.003$ m^2. The work of fracture would therefore be given by $1.903/0.003$ J/m^2 $= 634.4$ J/m^2. This number is actually much higher than the real work of fracture of glass, but it is only a hypothetical experiment.

Energy is the potential to do work, and therefore it has the same units as work. When a 2 kg brick is raised through 1 m it has the potential to do $2 \times 9.81 \times 1$ J of work, so its energy is also $2 \times 9.81 \times 1$ J. When a brick falls onto a piece of glass, supplying enough energy to equal its work of fracture, the glass breaks. Whereas tensile and compressive strength is a measure of how strong a material is, the work of fracture is a measure of how well the material resists being broken, a quality referred to as *toughness*. Materials that have a high work of fracture are tougher than those with low values. The work of fracture of glass is only 1–10 J/m^2, compared with approximately 1,000 for nylon, 10,000 for wood and 10^5–10^6 J/m^2 for mild steel.

Strain Energy

Energy is always required to break materials – to pull their atoms apart. But where does the energy come from to break samples in test machines like the Instron? The energy comes from the material itself, due to its having been stretched. This form of energy, which results from the strain in the material, is called *strain energy*. If you stretch an elastic band, you store strain energy in the rubber. If you suddenly let it go, the energy is released, and the work done on the band causes acceleration and a nasty sting when the speeding band hits your fingers. How can the amount of strain energy in a piece of stretched material be measured? If you had a spring balance you could attach a length of rubber to it, pull down on the rubber, and see what force (weight) was registered on the balance. The only other data you would need to calculate the strain energy stored in the rubber are the original (unstretched) length of the band, its dimensions, and the extension (Fig. 4.1). Suppose the force on the stretched rubber is F newtons. If you started releasing the tension, the reading on the spring balance would fall, until the reading was zero. The average force on the rubber, from the fully extended situation to the unstretched one, is $1/2$ F newtons. When the rubber is stretched, the force applied is moved over a distance equal to the extension. The work done in extending the rubber is therefore $1/2$ $F \times$ extension (joules). Therefore, the strain energy stored in the rubber is $1/2$ $F \times$ extension (J). This amount of strain energy is stored in the volume of the rubber and has to be expressed in joules per cubic meter. The strain energy in the rubber is therefore

$$1/2 = \frac{F \times \text{extension}}{\text{volume of rubber}} \quad \text{J/m}^3$$

The volume of the rubber band is the product of its cross-sectional area (*CSA*) and original length (l_o). Therefore, the strain energy is

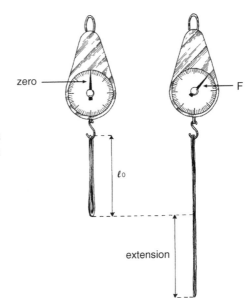

Figure 4.1. When a rubber band is stretched, strain energy is stored within the material, as described in the text.

$$1/2 = \frac{F \times \text{extension}}{CSA \times l_o} \quad \text{J/m}^3$$

But F/CSA is stress, and extension/l_o is strain. Therefore, the strain energy stored in a stretched material is 1/2 stress × strain.

Strain energy, as we will see later, has considerable functional importance during animal locomotion. It has a range of applications in our mechanical world, too.

Although it is useful to know that strain energy is half the product of stress and strain, a more useful relationship can be derived with a little mathematical sleight of hand. The expression 1/2 stress × strain can be rewritten as 1/2 stress × strain × stress/stress. Here, all I have done is multiply the original equation by stress/stress, and since stress/stress = 1, this has not changed anything. The equation therefore becomes strain energy = 1/2 stress × stress × strain/stress; therefore, strain energy = 1/2 stress^2 × strain/stress. Since strain/stress = 1/E, this relationship can be rewritten:

$$\text{Strain energy} = 1/2 \ \frac{\text{stress}^2}{E}$$

The strain energy stored in a material under stress is therefore directly proportional to the square of the stress and inversely proportional to its Young's modulus. This means that materials with lower Young's moduli store larger amounts of strain energy for a given stress. This conclusion explains why we youngsters used to employ rubber in our catapults and model aircraft. Materials and structures that have the capacity to store large amounts of strain energy may be referred to as being resilient. J. E. Gordon (1978, p. 90) defines *resilience* as "the amount of strain energy that can be stored in a structure without causing permanent damage to it."

Rudy Zimmermann was a World War II fighter pilot. Leaving Germany at the end of the war, he emigrated to Canada. His last job before retiring was with our Department of Vertebrate Palaeontology as a machinist. Rudy liked a good chat, and he told me all manner of things about aircraft and flying. One of his flying stories has to do with strain energy. The Treaty of Versailles, signed after World War I, prohibited Germany from having an air force, or flying powered aircraft, so young pilots learned to fly in gliders. Powered launching winches were few and far between, so the gliders were catapulted into the air. A long rubber cable was made into a V and attached, at its midpoint, to the towing hook of the glider. An equal number of fit young men would take a firm grip on either end of the cable while others held on to the glider. The two cable teams would then start running as hard as they could, leaving the glider behind them. When the cable was sufficiently taut, the third group would let go of their charge. The strain energy stored in the rubber cable was sufficient to accelerate the glider, with its pilot, to its takeoff speed.

Although wood has a much higher Young's modulus than rubber, it can be fashioned into a bow that can be used to store large amounts of strain energy when the string is drawn back. English archers of the Middle Ages used a longbow, which was drawn back by hand. Their enemies across the English Channel preferred the crossbow. Being considerably shorter than the longbow, the crossbow could not be drawn back by hand and had to be cranked back by a mechanical device. All this cranking took extra time, giving the English archers the advantage of a much higher firing rate. This advantage was the reason for the eventual decline of the crossbow on the battlefield, even though it continued to be used as a hunting weapon.

Since strain energy is directly proportional to stress squared, systems involving large stresses have the potential for storing large amounts of strain energy, even when the Young's modulus of the material is high. For example, when steel cables are being used to haul trawl nets, or for towing ships in distress, enormous stresses are involved. Sometimes cables snap, and the energy released can be sufficiently high to whip the broken cable across the deck with enough force to kill a person.

Windup clocks and toy trains are devices that utilize strain energy. Here the energy is stored in a coiled steel spring.

Energy Absorption

The other side of the coin of storing strain energy is that of absorbing strain energy, as when a mattress is used as a shock absorber for high jumpers. The absorption of strain energy, like its storage, is enhanced by low Young's moduli, as illustrated by the evolution of automobile tires. When the automobile replaced the horse-drawn carriage, iron-belted wagon wheels were replaced by wooden wheels with solid rubber tires. Since rubber has a low

Young's modulus, the tires absorbed much of the energy of the wheels moving over the ground, but the occupants still had a fairly rough ride, especially on the bumpy roads they had to travel. The invention of the pneumatic tire was a considerable improvement over the solid tire because compressed air has a much lower Young's modulus than rubber and therefore absorbs far more energy.

Evaluating Work of Fracture

The work of fracture of materials can be deduced from the data obtained during their tensile testing. Given that the Instron overestimates strain, the results obtained with it will not be accurate, but a worked example will show how you can use the data to evaluate work of fracture. To save possible confusion, you should remember that strain energy has the units of joules per unit volume (J/m^3), whereas work of fracture has the units of J/m^2.

During tensile testing on a standard test piece of mild steel, we obtained the following results:

Breaking load 495 lb = $495 \times 0.45 \times 9.81$ N.
Original length 51.7 mm = 0.0517 m.
Cross-sectional dimensions 7.6×1.0 mm; area = 0.0000076 m^2.
Extension at breaking point 10.54 mm.
Breaking stress = $495 \times 0.45 \times 9.81/0.0000076 = 2185.2/0.0000076$
     = 287.5 MPa.
Breaking strain = 10.54/51.7 = 0.204.

Recall that strain energy = 1/2 stress $\times$ strain J/m^3. (In practice this equation is closer to $1 \times$ stress $\times$ strain because the area under the stress/strain graph for steel under tension is closer to a rectangle than a triangle, but I will leave it as 1/2 here.)

Strain energy = $1/2 \times 287 \times 0.204 = 29.3$ MPa

This result is in joules per cubic meter, but we do not have a cubic meter; we have only 0.0000076×0.0517 m^3 = 0.000000392 m^3. Therefore, the strain energy in the test piece is:

$29.3 \times 10^6 \times 0.000000392$ J = 11.47 J

This acts over a cross-sectional area of 0.0000076 m^2. Therefore,

$$\text{Work of fracture} = \frac{11.47}{0.0000076} \ J/m^2 = 1.51 \times 10^6$$

The published values for the work of fracture of mild steel vary from 10^5 to 10^6 J/m^2 (Gordon 1978), so this result is not too far off.

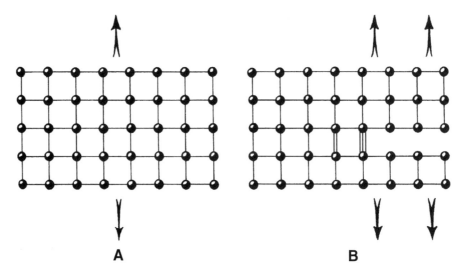

Figure 4.2. The tensile forces acting on a piece of material, tending to pull it apart, are resisted by the forces of the interatomic bonds, represented by the single lines (A). If there is a crack in the material, the adjacent interatomic bonds are subjected to larger forces, represented by double and triple lines (B).

What Happens When Things Break

When a piece of material is under tensile stress, the interatomic bonds are being pulled apart. Each of the bonds bears its own share of the tensile force. Provided this force is not too high, the interatomic bonds withstand the forces exerted upon them, and the material remains intact (Fig. 4.2A). However, if a crack exists in the material, those atoms whose bonds have been severed by the crack are unable to bear their share of the tensile load. Consequently, the adjacent bonds have to bear additional shares of the load (Fig. 4.2B). If the crack is not very long, the additional burden on the adjacent bonds is not too great, and they are able to tolerate the load without breaking. Obviously, the longer the crack the greater the share of the tensile stress the adjacent bonds have to bear. Eventually a point is reached when the additional load is too much to bear, and the overstressed bonds break. Their share of the load is therefore passed on to their neighbors, but they are also at their limit, and they break, too. This snowballing effect causes the crack to spread rapidly across the whole material, at the speed of sound, and the structure fails catastrophically. The crack length that sets this chain of events in motion, causing the material to fail, is described as the *critical crack length, L_g*. (The g stands for the engineer A. A. Griffith (1883–1963), who pioneered research on the subject.)

The phenomenon of critical crack length has applications in everyday life, but we are not always aware of it. Dressmakers and haberdashers make use of it when they tear off a length of cloth from a bolt of fabric. Using scissors they cut a small nick in the edge of the cloth – beyond the critical crack length – and then proceed to tear the material right across its width. People make use of the same phenomenon when they cut sheets of glass. Using a glass cutter, you first scribe a line across the piece of glass to be cut. (The term glass cutter is a bit of a misnomer because these tools do not cut through the glass but rather cut into the glass – to a depth of less than 1 mm.) You then line up the scribed line with the edge of the table and apply downward pressure to the glass. The pressure applies a tensile stress across the crack, and the glass breaks cleanly along the line. Glass tubing and glass rod is cut in a similar fashion in the laboratory. The piece to be cut is nicked, using a file or hacksaw blade, and pressure is applied to open up the crack. Those people who have tried laying bathroom tiles will also be well acquainted with this method. The tile shop usually supplies a tile cutter, a device that draws a cutting blade across the tile to be cut, thereby scribing the glazed surface with a shallow groove. Keeping the glazed side up, you then place the tile on a table and line it up so that the scribe mark extends slightly over the edge. By pressing down on both sides of the mark, you can break the tile along the desired line. I must confess that I have not enjoyed great success with cutting tiles, but that is due more to a lack of practice than to any problem with critical crack length.

The term critical crack length can be misleading because it is the depth of the crack as it penetrates the material, not its extent along the surface of the material, that is critical. To avoid confusion, think of the way the structure is being loaded relative to the crack. If the tensile force is tending to open up the crack, the length of the crack is measured in the direction of this extension in its depth. Suppose you held a ruler in your hand and flexed it sideways, placing one edge in tension and the other in compression (Fig. 4.3A). If there were a crack in the ruler, in the tensile edge, it would tend to open up with increasing

Figure 4.3. If a crack occurred in one edge of a ruler that was being flexed sideways (A), its length would be given by 1. If a ruler that was being bowed had a crack running across its width, the length of the crack would really be its depth (B).

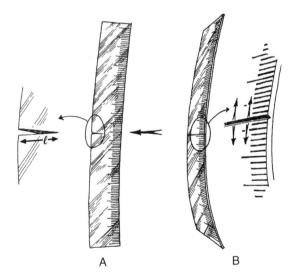

A   B

tension. The critical length of the crack would therefore be measured as shown in the figure, from the tensile edge toward the other. If a second ruler were taken and flexed so as to bow it, thereby bringing the two ends closer together, the concave surface would be in compression, the convex surface in tension (Fig. 4.3B). Suppose a crack ran across the width of the ruler on the tensile surface. In this case the critical crack length would be measured along the depth of the crack as it penetrated the surface of the ruler, not along the width of the ruler.

As the tensile stress on a material increases, so do the forces on the inter-atomic bonds; the critical crack length then decreases accordingly. It has been found that the critical crack length, L_g, is inversely proportional to the square of the tensile stress acting on the structure: $L_g \propto 1/s^2$. Critical crack length also varies with the toughness of the material. Tough materials, like steel, have longer critical crack lengths than materials, like glass, that are not tough. Hence $L_g \propto W$, where W = work of fracture. Lastly, stiff materials have longer critical crack lengths than those with low moduli, so $L_g \propto E$. Rubber, for example, which has a low Young's modulus, has a short critical crack length. This fact explains why the merest contact with a pin is enough to burst a balloon.

Why Things Break

Engineers can design structures whose materials will not be loaded beyond their tensile or compressive strengths nor energized beyond their work of fracture, but still structures sometimes fail. Aircraft crash, oil tankers break up at sea, and buildings collapse on land. Why? It seems the answer has to do with cracks, cracks that start off being shorter than the critical length but grow to reach dangerous lengths. Nothing much can be done to prevent cracks in materials – surfaces inevitably get scratched, and these scratches can grow into cracks. A simple experiment demonstrates how superficial scratches can affect the strengths of materials. Take new microscope slides and test them in a suitable machine until they break. Even though they are new, they have microscopic cracks (scratches) in their surfaces. Then repeat the experiment with another glass slide from the same batch, but this time treat the surface with hydrofluoric acid before testing. This acid dissolves glass, and the treatment therefore etches the top layer, removing the superficial scratches. The etched slides do not break as easily as the others, showing the effects of superficial scratches. Even if it were feasible to etch the surfaces of structures to remove superficial blemishes, it would be impossible to keep them scratch-free in everyday usage. Small cracks also occur within the material itself, but like the superficial cracks, there is nothing that can be done about them. Small cracks are a fact of life, and we have to learn to live with them.

The problem that small cracks pose is the way that stresses are concentrated at their tips. Suppose there is a small crack in a piece of material being loaded in tension (Fig. 4.4). The lines shown in the illustration are *stress trajectories,* which are like contour lines on a map. The closer together they are, the greater the local stresses. The crowding together of stress trajectories at the tip

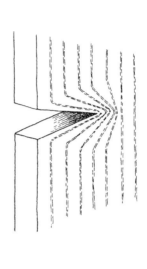

Figure 4.4. A crack in a piece of material that is being loaded in tension concentrates stresses toward the tip, as shown by the crowding of the stress trajectories.

of the crack depict the way that stresses are concentrated locally. The sharper the tip, the greater the stress concentration, which is expressed by the relationship $S \propto 1/r$, where S = stress at the tip of the crack and r = radius of curvature of the tip. The implication of this relationship is that blunting the crack, by increasing the radius of curvature of its tip, reduces the stress concentration, thereby preventing its spreading any further. This strategy certainly works, and it has practical applications. For example, you may have seen it used on the pockets of those inexpensive plastic raincoats, where a slit on the inside of the pocket allows you to reach inside your other pockets. To prevent the slit from getting longer, the manufacturer terminates it in a round cutout (Fig. 4.5A). When we go into the field collecting fossils, part of our equipment is a portable gasoline-powered rocksaw. The more expensive blades are made of steel, edged with diamond chips. The diamonds are inserted in slots, cut into the edge of the blade (Fig. 4.5B). To reduce the stress concentration at the bottom of the slots, each one terminates in a round hole.

Crack Prevention and Crack Stopping

The risk of cracks appearing in structures can be reduced by avoiding sharp corners in openings. For example, the windows in aircraft are round rather than square, and the doors have rounded corners rather than square ones. (Remarkably, the windows in the first Comet aircraft were square rather than round, and were probably the site of the failure of the fuselage.) Airframes (the skeleton of the aircraft) are subjected to large stresses, and certain key areas, such as engine pylons, are regularly inspected for cracks. One method used for inspection is to spray the structure with a dye that fluoresces in ultraviolet light and then to wipe off excess dye from the surface immediately after application.

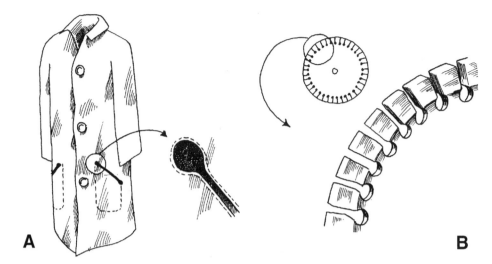

A   **B**

Figure 4.5. Plastic raincoats (A) and diamond rock saws (B) use cutouts to reduce stress concentrations.

If there are any cracks, they take up the dye by capillarity and therefore glow in the dark when illuminated with ultraviolet light.

Some years ago there was a series of disasters involving the McDonnell-Douglas DC-10. The problem was finally identified as cracks in the engine pylon, the strut that suspends the engine from the wing. The pylon was designed to withstand the bending forces exerted by the engine when running at full thrust, and should therefore not have failed. However, the maintenance workers of some of the airlines had weakened the pylons of the DC-10s they serviced by the unauthorized use of forklift trucks for removing engines during engine replacements. The excessive bending stresses exerted on the pylon by the forklifts resulted in small cracks. These cracks elongated during service, and some of them eventually reached the critical length. At this point the pylon broke and the engine tore away from the wing, causing catastrophic damage that resulted in the crash of the aircraft. Once the cause was determined, a directive was sent to all DC-10 operators to inspect the engine pylons for cracks, replacing any defective pylons. Since that time the DC-10 has had a long and safe flying record.

Structures, then, can be designed to minimize the chances of cracks arising, and critical structures can be regularly inspected for cracks. However, there is little that can be done to stop a crack from spreading once it has formed, except by changing the nature of the material. That can be done by using composite materials.

Composite Materials

The Egyptian Department of the Royal Ontario Museum has been excavating in the Dakhleh Oasis for many years, investigating Egyptian civilization during the period of the Roman occupation. The nearby rocks, which are Cretaceous in age, contain the fossil remains of extinct reptiles, so I went along with a paleontological colleague to do some prospecting. The fossil pickings were slim that year, but the archaeological sites I saw more than compensated for that shortcoming. On one occasion I accompanied the archaeologists for a day while they were working on a temple. Imagine walking out into the desert and seeing a temple that has stood there for 2,000 years, a temple made of mud bricks. Close inspection reveals that the bricks are composites of mud and straw, a building material that has been in use for thousands of years. This material is suitable only for dry climates, and a substantial period of rain would have collapsed the temple into a soggy mire. Aside from that limitation, mud, or adobe bricks as they are called, are very successful. Their toughness, that is, their ability to tolerate cracks without failing, is due to the inclusion of straw. When a crack appears, and many of the bricks strewn around the temple were cracked, it does not grow very long before it intersects a straw. The straw blunts the tip of the crack because it occupies a tunnel in the hard mud whose radius of curvature is much larger than that of the tip of the crack. As the straws are arranged randomly, no crack can go very far before meeting a straw, so none of the cracks can grow large enough to cause structural failure of the brick.

There are a wide variety of composite materials available today, and these, like the adobe brick, are made up of fibers embedded in a hard matrix. For example, when we collect fossils in the field we use a simple composite material for packaging the specimens that dates back to the pioneer dinosaur collecting days of the last century. The fossil, still partially embedded in a block of rock, is first covered with wet paper to act as a separator. Then strips of burlap are dipped in wet plaster and wrapped around the block. Two layers of the burlap-plaster composite are usually used, and they take only a few minutes to set. Once the field jackets have thoroughly hardened, they are remarkably tough and can be tumbled and manhandled over the rocks without damage.

Fiberglass, made from embedding glass fibers in polyester or other resin, is a more recent composite. It has been in wide use since the 1950s for a variety of applications, from building car bodies and small boats to making repairs. It is very tough, fairly light, and easy to use, so it has been popular with do-it-yourself enthusiasts. Another kind of composite, which uses carbon fibers embedded in a resin, has been used in recent years for making tough lightweight tennis rackets.

Reinforced concrete, made by embedding steel tendons in concrete, may be thought of as a composite material. However, the fibers are considerably coarser than they are in composites like fiberglass, and the strength of this material probably has less to do with crack-stopping than it does with combining the compressive strength of concrete with the tensile strength of steel. Bone, which is a composite of resilient collagen fibrils and stiff calcium phosphate crystals, is sometimes compared to reinforced concrete. However, as we will see in the next chapter, the relationship between the two components of bone is far more intimate than it is in reinforced concrete, and its crack-stopping mechanisms are a major contributing factor to its toughness.

PRACTICAL

Measuring the work of fracture of a material is not easy: "[T]he measured value of the work of fracture does depend on the way in which the test is done, and it is difficult to get consistent figures . . ." (Gordon 1978, p. 105).

We saw earlier how approximate values for the work of fracture can be obtained from Instron results. We could also get some direct comparative values using a simple ballistic apparatus (Fig. 4.6). With this apparatus we would drop a weight from a known height onto a cutting die that rests on the test piece. We could calculate the amount of energy the die supplied to the test piece simply by measuring the distance through which the weight fell and multiplying this measurement by the mass of the weight (work = force × distance; force = mass × acceleration; work is in joules, force is in newtons, and distance is in meters). We would then repeat the experiment with increasing amounts of energy (greater height and/or greater mass) until the test piece breaks. The experimental design is rather crude, and there are several sources of error, including the following:

1. Some of the energy of the falling weight is absorbed by the resilience of the test piece, by the table, by the metal stand, and also in the transfer of energy from the weight to the cutting die.
2. It is virtually impossible to get the falling weight to transfer exactly the same amount of energy each time.
3. Once the test material has been struck, its surface is usually scored, which weakens the entire test piece.

In order to express the work being done on the cross-sectional area of the material, we would have to take into account the turning moments (turning effects, see p. 123 for explanation) set up in the material at the point where it fractures. It would not be difficult to work this factor into the calculation of the results, but the imprecision of the apparatus hardly warrants the extra effort. Because of these problems the apparatus is used only to make comparative tests, with no attempts to try to calculate the work of fracture. Results are therefore expressed in terms of the amount of work required, in joules, to break the test sample.

The results are quite comparable because, with the exception of the glass microscope slides, the test samples are the same size (there are slight variations but they can be ignored). The glass slides are about one-sixth thinner than the other test pieces, therefore it is necessary to multiply the amount of work to break the slides by *six* to get comparable results. To help overcome the problem of weakening the test piece by repeated trials, you are given the approximate values of the weight and the distance from which it needs to be dropped to break the material. Assume that the weights have masses of 50, 100, 200, and 1,100 g.

Experiment 1   Assess the amount of work required to break a microscope slide.

CAUTION: Goggles to be worn for this experiment.

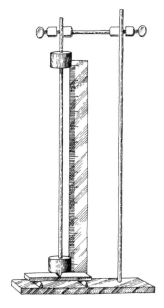

Figure 4.6. A simple apparatus for quantifying the work of fracture.

Place the slide on the stand so that there is an equal overhang at each end. Load a 100 g weight onto the rod, then place the lower end of the rod into the hole in the cutting die. Place the cutting edge of the die onto the glass slide, making sure that:

1. It rests in the center of the slide.
2. It lies at right angles to the length of the slide.
3. The rod is vertical.
4. The weight will clear the ruler.
5. You remember your goggles.

Lightly holding the cutting die to steady it, raise the weight to the desired height. (The slide will probably not break in a distance of less that 20 mm.) Prior to releasing the weight, remove your hand from the cutting die, otherwise you may absorb some of the energy. Repeat, increasing the distance in 2 mm intervals, until the sample breaks.

Worked Example:
Weight used = 100 g
Distance at which sample broke = 32 mm.
Work done = $0.1(kg) \times 0.032(m) \times 9.81(m/s/s) = 0.0314$ joules.
Multiplying the result by 6 to make it comparable with other test pieces = $0.0314 \times 6 = 0.188$ joules.

Repeat the experiment with a second microscope slide, making sure to wipe the cutting edge and the two metal supports (otherwise you may get stress concentrations that will cause premature breaking). If supplies of glass slides hold out, you may like to try five or six times.

How repeatable is the experiment? I obtained a fairly wide range of breaking distances. I am inclined to attribute this to the imprecision of the test rather than to variation in toughness of the glass from one slide to another.

Notice the shape of the broken halves. They fit together precisely, which is what you would expect of a brittle material (no plastic phase). Notice also the cracks in the piece of glass that gets cut from the middle of the slide.

Before commencing Experiment 2, you have to prepare some test samples. You are provided with a silicone rubber mold of the standard test piece, some polyester resin, a catalyst, a woven glass fiber mat, paper cups, wooden stirring rods, and a brush. Mix a small quantity of resin with a few drops of catalyst in a paper cup. Once it is stirred, pour some of the resin into the mold until it just reaches the top. The catalyzed resin will take about thirty minutes to set, after which time you can remove it from the mold. While it is setting, make a second cast, this time from laminated glass fibers and resin. Start by cutting several pieces of glass fiber mat to the exact size of the mold (check to make sure they will fit snugly). Examine the mat and see that it is made up of woven fibers of glass. Try not to get the fibers on your person; they are a strong irritant. Using the brush provided, paint a layer of catalyzed resin onto the bottom and sides of the mold. Drop in a piece of the precut glass mat and tamp it down well with the brush to ensure that the resin completely saturates the fibers. Paint a layer of catalyzed resin on top of the glass mat, then add another piece of mat. Continue the process until you have made a laminated cast of the test piece. Allow thirty minutes setting time, and remove the cast from the mold. Take care not to cut yourselves on any sharp edges when handling the laminated test piece.

Experiment 2   Determine the amount of work required to break polyester resin test pieces.

Set up the pure resin test piece in the impact apparatus, as you did for the glass slides. Determine the energy required to break it (guide: 200 g weight at about 150 mm). Did you see any obvious cracks in the material prior to breaking? Probably not.

Repeat with the laminated sample. You will find that you have a good flat side, which was at the bottom of the mold, and a rougher top side. Place the sample with the good side down. Determine the energy required to break the test piece (guide 1,110 g at 65 cm – you may not be able to break it!).

Notice how prominent cracks form in the material long before the breaking point is reached. These cracks do not spread right across the sample as soon as they are formed because the glass fibers absorb the energy and the crack fails to propagate.

The Storage of Strain Energy   We have seen that the amount of strain energy that can be stored in a piece of material is inversely proportional to its Young's modulus. Devices used to store strain energy therefore often employ materials of low moduli, such as rubber. Examine the different ballistic devices provided (Fig.

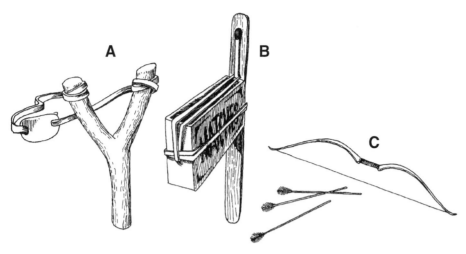

Figure 4.7. Some simple ballistic devices that use stored strain energy for propulsion: (A) catapult; (B) matchbox gun; (C) bow.

4.7). Some may even evoke memories of your ill-spent youth. You may want to try making one of your own with elastic bands, a matchbox, and a lollipop stick, *but be very careful of all ballistic devices – even the small ones can be dangerous.*

The Absorption of Energy If two materials are subjected to the same stresses, their respective abilities to absorb energy, like their abilities to store strain energy, are inversely proportional to their Young's modulus. You can test this statement for yourself by comparing the relative performance of different materials in protecting a glass slide from being broken in the impact apparatus.

Experiment 3 Compare the energy absorption performances of materials of different modulus.

Set up the impact apparatus to break a glass microscope slide, as previously. However, instead of placing the die directly onto the slide, place it onto a pad of material that will absorb some of the energy of impact. You are provided with standard-sized pads of rubber, paper, and wood. Determine whether their capacity to absorb energy accords with their Young's modulus.

As an aside, you can test the efficacy of even small quantities of materials of low modulus to absorb energy. Try placing one, then two, sheets of paper onto the surface of a slide prior to breaking it. Is the amount of energy absorbed by the paper significant? Can you think of any biological examples of energy absorption?

5 Bone as a Composite Material

I have a sheep rib in front of me. It is not very large – about as long as a pencil – and although somewhat flatter and wider, it has about the same amount of material as a pencil. I would have no difficulty breaking a pencil, and the rib looks no harder to break. I take a firm grip with both hands and start applying pressure, gently at first because I am sure it will break quite easily. But it does not break. I apply a lot more pressure and can feel the bone giving a little – it is elastic like the blade of a sword – but it will not yield. It must break soon. I am really trying my hardest to break the rib now, but it refuses to cooperate. I have to admit defeat and put the rib down. I have a pair of depressions in the palms of my hand where the ends of the rib dug into me, but the rib itself is unscathed. No cracks, no bends, nothing. Its behavior was elastic throughout.

I have two other sheep ribs in front of me. They look the same as the first rib, though one of them looks a bit brown compared with the others. This is because I baked it in the oven for a couple of hours, at about 150° C. The other rib was placed in 10 percent mineral acid for several days. It looks just the same as the untreated rib, the same color and the same shape, but when I pick it up the differences between them are immediately apparent. It is soft and flexible like a piece of rubber. I can even tie it into a knot! Treating it with acid has dissolved away the mineral portion of the bone, the calcium phosphate, leaving the collagen. *Collagen,* a protein, is a rather tough connective tissue that is found in many parts of the body including the skin, the walls of blood vessels, and tendons and ligaments. *Tendons* are the tough sinews that connect muscles to bones, and *ligaments,* which are similar to tendons, connect bones together. Incidentally, when collagen is boiled for a long time the collagen hydrolyzes, becoming converted into gelatin, which is used for making jelly, among other things. The decalcified bone is compliant (the converse of stiff) and would not break if dropped onto a ceramic floor. But it is not very tough, and I could pull it apart quite easily.

Most proteins are denatured when heated to about 100° C, and baking the second rib broke down its collagen. This made the bone brittle, and I can break pieces off with my thumbnail. Only slight pressure is required to snap the rib, and if I dropped it onto the floor it would shatter. However, the bone is still stiff, even though it is no longer tough.

Figure 5.1. A broken rib showing the outer layer of compact bone and the spongy interior of cancellous bone.

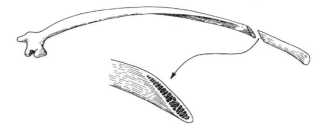

Neither collagen nor the mineral portion of bone is tough on its own, but when they are combined, the resulting composite is remarkably tough and compliant.

If I break the brittle rib in half and examine one of the broken ends, I can see that the bone comprises an outer layer, the *compact* bone, and a spongy interior, called *cancellous* bone (Fig. 5.1). The functional relationship between compact bone and cancellous bone will be discussed in the next chapter; our concern here will be with compact bone. It is from compact bone that test samples are taken for mechanical testing, and most of these tests have been performed on human bone.

Haversian Bone

Human compact bone, like that of most mammals, birds, and some extinct forms including many dinosaurs and pterosaurs, has a characteristic microscopic appearance (Fig. 5.2A–G). This type of bone is described as Haversian, named after Clopton Havers (1655–1702). Most textbook accounts of skeletal materials describe Haversian bone, and it is easy to get the impression that it is the typical bone of vertebrates, but that is not so. The bony fishes, which account for more species than any other vertebrate group, do not have Haversian bone, nor do amphibians nor most reptiles. Mammalian bone is not exclusively Haversian either, since other types occur as well. Haversian bone is therefore atypical of vertebrates as a whole.

Haversian bone is characterized by the regular arrangement of the bony matrix as a series of concentric lamellae, like the layers of an onion or leek. The concentric lamellae surround a central longitudinal canal called the Haversian canal. Although these canals run essentially parallel to the long axis of the bone, they meander and fork and should therefore be visualized as branches of a tree rather than as drinking straws in a box. They are also cross-connected, at intervals, by horizontal channels called Volkmann's canals. The central canal is occupied by a capillary and a nerve; the former is for supplying nutrients to the living bone tissue, but the function of the latter is not understood. The bone cells, called *osteocytes* (Greek *osteon*, bone; Greek *kytos*, a hollow vessel, now often used to mean a cell), lie within a cavity, the *lacuna* (Latin *lacuna*, a pit). The lacunae are about 25 micrometers across (a micrometer, abbreviated μm, is 10^{-6} m). Fine fluid-filled canals called *canaliculi* (Latin

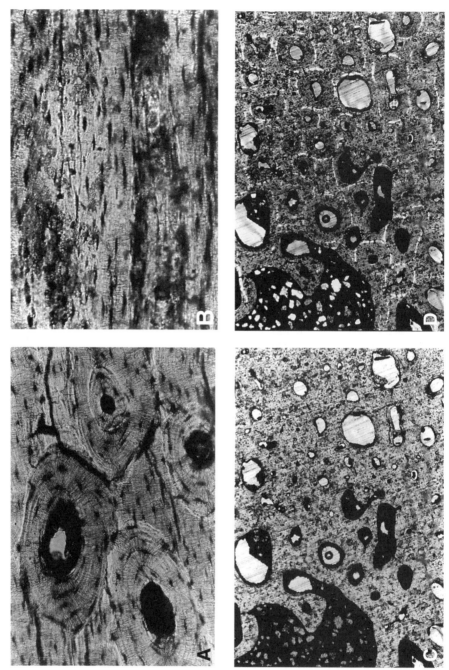

Figure 5.2. Photomicrographs of bone and teeth. The magnifications given are for comparative purposes only, and refer to the magnification of the microscope when the photographs were taken. Because of photographic enlargement, the actual magnifications are about n times higher than the magnifications as given, where n = length of photograph (mm)/35. (A) mammalian bone, transverse section, ×200; (B) human long bone, longitudinal section, ×200; (C) the dinosaur *Ornithomimus*, transverse section through a phalanx, ×40; (D) the same, under polarized light.

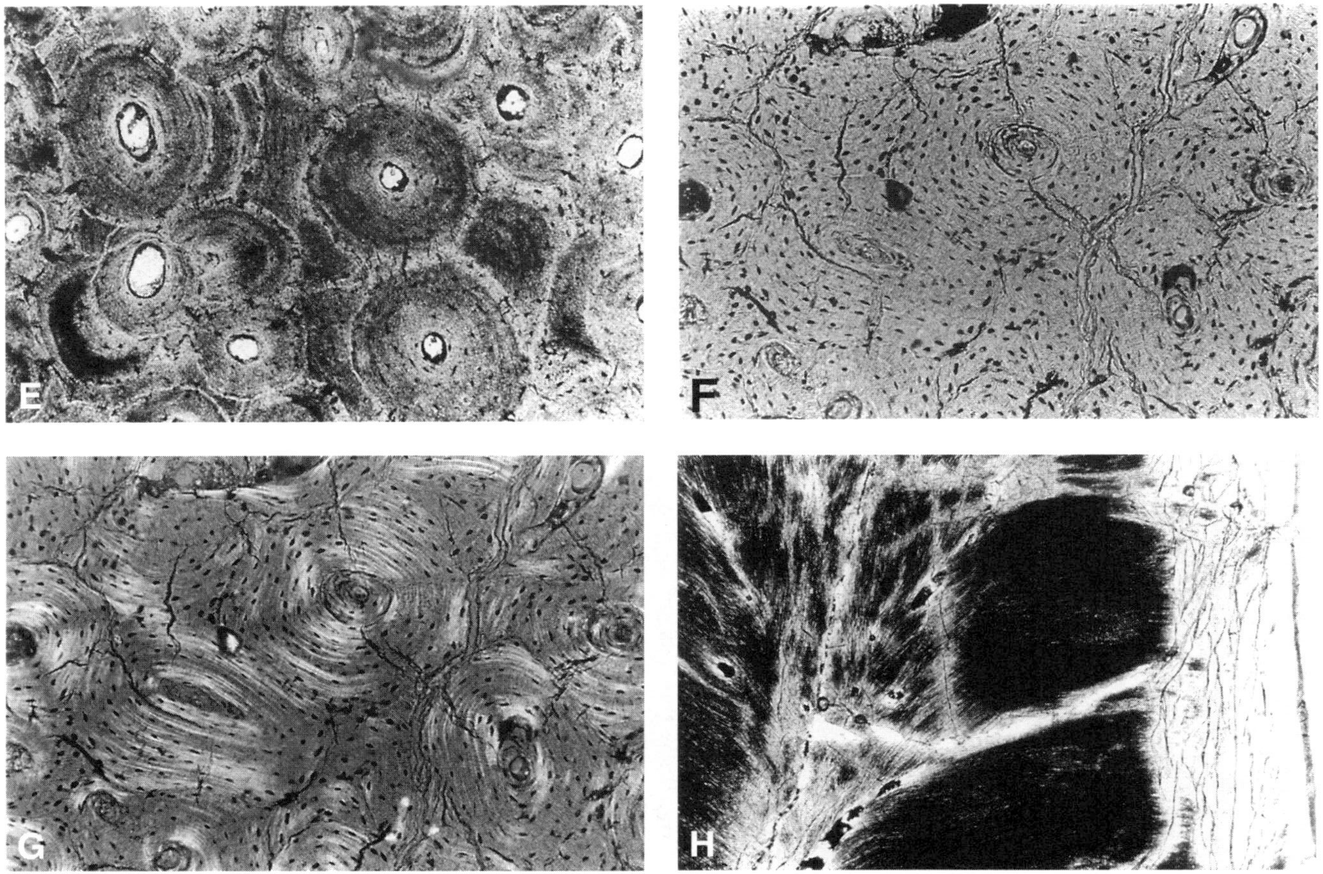

Figure 5.2. *(cont.)* (E) the horned dinosaur *Chasmosaurus belli*, transverse section through a rib, ×100; (F) human long bone, transverse section, ×100; (G) the same, under polarized light; (H) a hadrosaurian dinosaur tooth, in longitudinal section, ×40.

canaliculus, small channel) radiate in all directions from the lacunae, conveying cytoplasmic extensions from the osteocytes. The canaliculi of adjacent lacunae are interconnected, and those of the innermost lamellae interconnect with the fluid-filled Haversian canal. Nutrients are therefore free to diffuse from the central blood vessel, via the surrounding fluid, to all osteocytes, via the anastomosing system of canaliculi. Similarly, waste materials diffuse in the opposite direction. Therefore, although the living bone cells are embedded in a hard, impervious matrix, they are in fluid continuity with the blood vascular system. Since none of the osteocytes are farther than about 0.15 mm from the nearest blood vessel, the rate of diffusion is sufficiently rapid to ensure their survival. The concentric arrangement of bone lamellae around a central canal is therefore a functional solution to the problem of supplying the living bone cells with their needs.

A single unit of Haversian bone, consisting of a Haversian canal with its surrounding lamellae of bone matrix, is called an *osteon* (sometimes spelled osteone). Haversian bone is described as being *secondary bone* because it replaces the previously existing *primary bone* formed during earlier development of the individual. Sometimes the primary bone that the Haversian bone replaces looks very much like Haversian bone. However, the *primary osteons* of young individuals are easily distinguished from the *secondary osteons* of Haversian bone because the latter have a cement layer around the outside (see Currey 1984, pp. 29–30). The secondary osteons are also larger than the primary osteons (200–300 µm diameter in humans). The cement layer, which is 1–5 µm thick, is not obvious when viewed under ordinary light (Fig. 5.2C). However, it shows up clearly under polarized light as a white zone, and it is referred to as the *cement line* (Fig. 5.2D). As we will see later, the cement line plays a critical role in crack-stopping. The cement is devoid of collagen and contains less calcium and less phosphorus than the bone lamellae and more sulphur. Its precise composition has yet to be established (Martin and Burr 1989). Cement appears to have a lower Young's modulus than that of bone matrix, and its behavior is more plastic.

Although bone may appear to be an inert material, we must remember that it is a dynamic living tissue and is continually being modified. Old osteons are broken down by specialized cells called *osteoclasts* and are then replaced by new ones. The remnants of old osteons can be seen interspersed between their replacements, and these partially resorbed osteons are called *interstitial lamellae* (Fig. 5.2E). When primary osteons are replaced by secondary ones, the bone actually becomes weaker, both in compression and in tension. Secondary bone also undergoes more plastic deformation than primary bone, probably because the cement line permits slippage to occur, which allows osteons to be pulled out from the rest of the bone in regions of peak stress, like drinking straws being pulled from a box of straws.

Collagen accounts for about one-third of the dry weight of bone, and the mineral portion accounts for almost all of the rest. The collagen is organized into fine fibers that are about 40–120 nm (nanometers, 10^{-9} m) in diameter (reported in Pritchard 1972). The bone mineral is a particular form of calcium phosphate called *calcium hydroxyapatite* [$Ca_{10}(Po_4)_6(OH)_2$]. The hydroxyapatite appears to be in the form of small crystals, only a few namometers across, but there is some difference of opinion as to whether they are needle-shaped, or platelike (Currey 1984, p. 26). These crystals are attached to the surface of the collagen fibers. The fibers are either arranged longitudinally, with their long axes parallel to the long axis of the osteon, or they are tightly spiraled, so that they lie at right angles to the long axis of the osteon. The collagen fibers in a particular lamella are predominantly oriented the same way, either longitudinally (type L of Martin and Burr 1989), or transversely (type T). The two arrangements have different appearances under polarized light, and, as we will see later, they also have different mechanical properties. The type L lamellae appear dark, and the type T appear light. Consequently, osteons composed entirely of type L fibers look uniformly dark under polarizing light, whereas those composed solely of type T fibers appear light. In some osteons, lamellae with type L fibers alternate with those of type T, so their lamellae alternate light and dark under polarizing light (Fig. 5.2G). These osteons are designated as type A (for alternating) and appear to be intermediate in mechanical properties between those of types L and T. The type T osteons are stronger in compression than those of type L (164 compared with 110 MPa), and they are also somewhat stronger in shear (57 compared with 46 MPa), but the reverse seems to be true in tension. (Shear, recall, is transverse movement, as when a pack of cards is pushed sideways, causing the cards to slip upon one another.) There are also differences in their Young's moduli: in compression, type T are marginally stiffer (9.3 GPa) than type L (6.3 GPa), but the reverse is true in tension (data reported in Martin and Burr 1989).

Since the orientation of collagen fibers is correlated with mechanical strength, we could expect to find differences in fiber orientation in different parts of a bone according to whether they are loaded in tension or compression. Are such correlations found? Portigliatti-Barbos, Blanco, Ascenzi, and Boyde (1984) claimed it to be true for the human femur. The anterolateral portion of the shaft is primarily loaded in tension and predominates in longitudinal fibers, whereas the posteromedial aspect, which is loaded in compression, has more transverse fibers. However, this study, and its results, have been criticized on several grounds (Riggs, Lanyon, and Boyde 1993). First, the human femur is probably subjected to a wide range of loading regimes and should therefore not be expected to experience a particular loading pattern in any given region. Second, many muscles attach to its shaft, and their contractions further complicate the local stress patterns in the bone. The study was also criticized for sampling too few bones. These shortcomings were overcome in a study on a large sample of the radius of the horse (Riggs, Lanyon, and Boyde 1993). The equine radius has a relatively simple loading regime that has been analyzed by several in vivo studies (meaning in the living body, as opposed to in vitro, meaning in glass, referring to the laboratory test tube). Its convex anterior aspect is always loaded in tension, whereas the concave posterior region is always loaded in compression. The absence of muscle attachments from most of its shaft contributes to the simplicity of its loading. The study showed that the osteons of the posterior (compressive) aspect predominate in transverse and oblique collagen, whereas the rest comprise mostly longitudinal fibers. Significantly, the matching of fiber orientation with loading regimes was not seen in the fetus, nor was it very apparent in foals, but it was seen in all adults of two years and older. In a follow-up study Riggs, Vaughan, Evans, Lanyon, and Boyde (1993) showed that bone from the anterior region (tensile) was significantly stronger in tension than bone from the posterior aspect (compressive), the respective values being 160 MPa and 104 MPa. Conversely, the values for compressive strength were 185 MPa and 217 MPa, respectively.

Crack-Stopping and Toughness of Haversian Bone

Haversian bone, as we have seen, is a composite material of collagen fibers and calcium hydroxyapatite crystals. Although this is its composite structure at the finest microscopic level, at the functional level, in terms of how bone breaks, the osteon is the fiber unit, and the cement layer that surrounds each osteon is the equivalent of the embedding matrix. The possible crack-stopping role of the discontinuities provided by the Haversian canals, and by the lacunae, has been discussed in the literature, and both may serve such a role. However, it seems that the primary crack-stopping mechanism involves the weak interface between the cement and the enclosed osteon. There are also weak interfaces between interstitial lamellae and the adjacent osteons, as well as between adjacent lamellae within a single osteon; these, too, contribute to crack-stopping. As we saw earlier, the presence of secondary osteons weakens bone, but the reduction in tensile and compressive strengths of secondary bone over primary bone is more than compensated for by the marked increase in toughness.

Breaks occur, recall, when small cracks appear within a material and are allowed to grow without interruption. Toughness can therefore be achieved either by reducing the formation of small cracks in the first place, or by stopping their growth when they do occur. Bone differs from nonliving composites in that it is continually being replaced, so damaged parts are eventually replaced by new bone. Consequently, the accumulation of small cracks in bone is of less significance than it is in other materials, and crack-stopping in bone is more relevant than reducing the incidence of cracks. When a crack appears, it typically elongates slowly at first, accelerating only prior to producing a catastrophic failure. If the forces acting on a bone are large enough, as when a speeding horse puts a foot wrong, or a downhill skier stumbles and twists her leg, a catastrophic failure is the inevitable outcome, regardless of the crack-stopping mechanism. However, when lesser forces are involved, the crack is arrested and its energy dissipated. Visualize a small crack – the microfractures

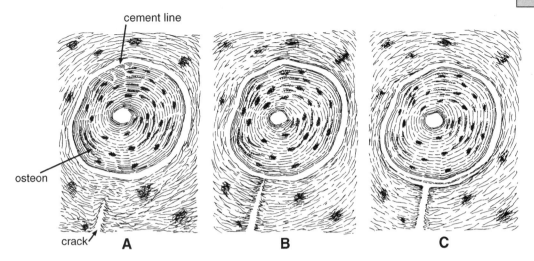

Figure 5.3. A small crack advances toward an osteon (A), enters the cement line (B), and thereby becomes blunt and stops (C).

mentioned in Chapter 3 – encountering a cement line. The tip of the crack is sharp, with a small radius of curvature, but when it enters the cement and the latter fails, it opens out into a larger crack, blunting its point and dissipating its energy (Fig. 5.3). Sometimes, as when a bone is loaded in bending, the forces on the tensile side cause osteons to be pulled out as the crack extends; this phenomenon has been shown experimentally (Fig. 5.4). The energy required to bring about shearing as the osteons (the fibers of the composite) are pulled out

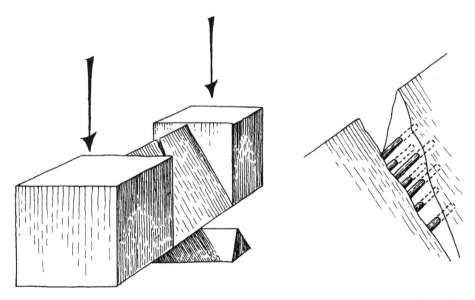

Figure 5.4. A bone sample, machined into a wedge and loaded in bending such that individual osteons are pulled out of the cement line. (After Martin and Burr 1989.)

of the cement (the matrix) robs the crack of energy, which may be enough to stop it. You can see this effect during microscopic examination of the fracture surfaces of bone samples broken during testing. The bone in the initial part of the crack appears ragged, where osteons have been plucked out. However, the rest of the fracture surface is cleanly broken, where the accelerated propagation of the crack has sliced right through the osteons (reported in Martin and Burr 1989).

The effective crack-stopping mechanisms of Haversian bone give it excellent fatigue-resisting properties, which is probably of considerable importance to active animals like cursors (running animals, see Chapter 10). Experiments in which bone was subjected to cyclic loading and unloading showed there was an initial decrease in Young's modulus, associated with the initiation of numerous cracks (reported in Martin and Burr 1989). Stiffness continued to decrease, but after about 3 million cycles Young's modulus remained unchanged. There appeared to be no reduction in bone strength when the samples were treated at strain rates comparable to those expected in life, and failure did not occur even after 30 million cycles. In human terms, 30 million cycles is roughly equivalent to the loading a femur would experience during twenty years of walking five miles a day. Given that the rate of replacement of compact bone in adult humans is about 5 percent per year, all of the bone material would have been replaced by the time the 30 million cycles had been reached. While the ongoing replacement of Haversian bone may repair the small cracks that accumulate, there is evidence that bone repair may also be initiated at the site of damage. Bone, then, combines the advantages of its inherent resistance to fatigue failure, attributed to its fine structure, with a built-in repair mechanism.

Bone, like wood, is markedly directional in the orientation of its fibers, the osteons running predominantly parallel to the longitudinal axis of the whole bone. This structure maximizes the strength of the bone in the direction of the predominant loads, but it does so at the expense of strength in shear and torsion (twisting). This facts helps explain why most accidents that result in fractures involve the bending or twisting of bones.

Variation in Mechanical Properties

There is not very much variation in the mechanical properties of compact bone among different mammals, nor among different bones within the skeleton. Values for Young's modulus, the work of fracture, and for the density of a sheep's femur would therefore be similar to those for a human femur. There would also be little difference between these values and those for the tibia or for a metatarsal bone. However, there are marked differences in the properties of bones that serve different mechanical roles. Currey (1984) compared the mechanical properties of bone from the femur of a cow (*Bos taurus*), the antler of a red deer (*Cervus elaphus*), and from the tympanic bulla of a fin whale

(*Balaenoptera physalus*). Before looking at the marked differences in their properties, we need to consider the ways in which these structures are used.

The femur is used for both weight support and locomotion, and therefore needs to be stiff, strong, and tough. Like other limb bones, the femur is accelerated, and retarded, during locomotion, and the energy required for this movement is directly proportional to the mass, hence to the density, of the bone. Minimal bone density optimizes energy conservation but decreases stiffness, so a compromise density is required that gives the bone sufficient stiffness.

Antlers, which are characteristic of deer (family Cervidae), are bony structures that are shed each year. Usually it is only the male that has antlers (*Rangifer,* the genus that includes the caribou, is an exception), and they are used in male-male competitive interactions and in attracting females. Each year the new set of antlers is larger and more branched than in the previous breeding season, so older males have an advantage over their younger rivals. Most of the interactions between rival males involve ritualized displays in which the size and spread of the antlers are used to impress the other. The antlers would not need to be tough or strong for this function, nor particularly stiff. However, the interactions often lead to fighting, where the antlers are banged together, becoming interlocked in grunting, heaving, head-pushing trials of strength. The antlers have to be tough and strong to withstand this kind of wear and tear.

The tympanic bulla is a bony capsule housing the ear apparatus. Directional hearing hinges on detecting slight differences in sound quality reaching the left and right ears. For example, when crossing a road, we know that the sound of an approaching car is coming from the left rather than from the right because the sound reaching the left ear is slightly louder. We may also be able to detect the slight difference in the arrival times of the sounds at the two ears. This mechanism works only because the sensory organs of the left and right ears are isolated from one another within their own bony capsules.

When sound waves strike an object they may be completely reflected, completely absorbed, or a combination of the two, depending on the differences in density and Young's modulus between the object and the medium in which the sound travels. The greater the differences, the greater the extent of reflection over absorption. Brick, for example, has a higher density and higher Young's modulus than plaster, which is why brick partition walls are more soundproof than plasterboard ones. Most of the sound waves that strike our heads are reflected away because of the great differences in density and modulus between air, flesh, and bone. The only waves reaching the sound receptors are those that pass down the bony canal running from the earflap to the eardrum. However, when we go swimming underwater, the density and modulus differences between our heads and the surrounding water is much less, so we essentially lose our directional hearing ability. You can check this out for yourself in the swimming pool. Cetaceans, in contrast, have excellent

Table 5.1 Physical Properties of Different Types of Bone

Property	Femur	Antler	Bulla
Young's modulus (GPa)	13.5	7.4	31.3
Bending strength (MPa)	247	179	33
Work of fracture (Jm^{-2})	1710	6190	200
Density (kgm^{-3})	2.06	1.86	2.47

Data from Curry 1984.

directional hearing underwater. Part of the reason for this directionality is that the bone of the tympanic bulla has such a high density and high Young's modulus that it maximizes the differential with water.

The mechanical properties of bone from femora, antlers, and tympanic bullae are shown in Table 5.1. The values for the femur may be considered as "standard bone," against which the others can be compared. Bending strength is like tensile strength, except that the test sample is broken in bending rather than in straight pulling. Thus *bending strength* may be defined as the maximum stress in the material immediately before it fails in bending. Antler, not surprisingly, has by far the highest work of fracture of the three types of bone, whereas that of the tympanic bulla is very low. The bulla does not need to be tough because an external force large enough to break it would be fatal to the individual anyway. Similarly, the bending strength of the bulla is very low, too. Antler is not quite as strong in bending as femur, suggesting that the stresses involved during ritualistic fighting are less than those experienced by limb bones during locomotion. Alternatively, this difference may reflect only that breaking an antler is less serious than breaking a leg. Bulla is by far the stiffest bone, maximizing the difference in modulus with water. Antler has the lowest Young's modulus, enhancing its ability to absorb strain energy and therefore to act as something of a shock absorber during head-butting activities. Predictably, bulla has the highest bone density, correlated with sound reflection, and antler has the lowest, correlated with its low Young's modulus.

Almost all of the discussion so far has pertained to adult bone, but I would like to close the chapter by making a few points about immature bone. The compact bone of young animals, as mentioned earlier, consists mostly of primary osteons and is stronger in tension and compression than that made up of secondary osteons. Young bone also has a lower Young's modulus, making it far more resilient than adult bone. Since it can absorb more strain energy than adult bone, young animals are less likely to break bones during falls than mature ones. Human toddlers, for example, seldom suffer more than cuts and grazes during their misadventures with gravity, whereas adults, under similar circumstances, are more likely to break bones. (The fact that toddlers also weigh much less than adults and have less distance to fall is probably relevant, too.) And when youngsters do load their bones beyond failure, they often bend rather than break, giving the characteristic "green-stick fracture." It has been suggested that the mechanical properties of young bone are an adaptation to differences in lifestyle – adventuresome youth against cautious adult. However, this is not necessarily so. The bone of a neonate axis deer (*Axis axis*), for example, has bone that is more similar to that of human adults than to that of children in its stiffness, due to its being well mineralized (Currey and Pond 1989). The adaptive significance of this stiffer bone is that most newborn ungulates have to be able to run with the herd soon after birth, so their skeletons have to be stiff.

PRACTICAL

You will be looking at some thin sections of bone under the microscope to familiarize yourself with histology (*histology* is the study of tissues). But before you do so, I want to review magnification factors and pass on a useful tip about microscopic scales.

The eyepieces and objective lenses of most microscopes have a magnification factor engraved on them. Objective lenses typically have magnifications of 4, 10, 20, 40, and 100 (oil immersion), whereas eyepieces are usually ×5 or ×10. The magnification of the image being examined is simply the product of the magnification factors of the eyepiece and the objective lens. If you are looking at a specimen using a ×5 eyepiece and ×20 objective lens, the image is 100 times larger than life. However, this knowledge does not give a scale whereby you can estimate the sizes of certain features of interest. Ruled graticules are available for this purpose, so, too, are eyepieces with built-in scales, but there is a much simpler and cheaper way of achieving this end.

Human red blood cells (erythrocytes) have an average diameter of 7.5 micrometers ($\mu m = 10^{-6}$ m) and make ideal rulers for measuring microscopic structures. All you need do is make a blood smear on a microscope slide. You need to spread the erythrocytes thinly enough that they do not lie on top of one another. You'll need two microscope slides, a small, sterile syringe needle (still in its sterile wrapper), a tissue, and some alcohol. Make sure the microscope slides are clean and grease-free. If in doubt, clean them off with alcohol. Leave both slides on the bench. Wet the tissue with alcohol and thoroughly wipe one of your fingertips. Swing your arm back and forth a few times to pool the blood in your fingers. Then, without stopping to think about it, jab the sterilized part of your finger with the sterilized needle and squeeze out a drop of blood. Gently touch the drop of blood onto one end of one of the microscope slides. Do not dab the finger against the glass – you want to create a droplet of blood on the slide, not a smudge. Holding the second microscope slide at an angle of about 45°, lower one end until it makes contact with the middle of the first slide, with its lower surface facing toward the blood droplet (Fig. 5.5). Slowly draw the inclined slide toward the blood droplet. When it touches it, the blood will flow along its entire width by capillarity. As soon as this happens,

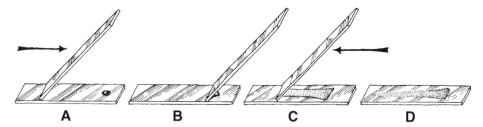

Figure 5.5. A blood smear is made by placing a drop of blood on a clean slide (A), drawing a second slide up to it until the droplet spreads along its width by capillarity (B), and pushing the second slide along the first, thereby drawing out a thin smear (C, D).

push the inclined slide rapidly back in the opposite direction, making sure to keep the same inclination, and firmly press the edge against the surface of the other slide. You should finish up with a thin film of blood. The erythrocytes will be more thinly spread out at the end farthest from where the droplet was deposited. Examine them under the microscope at different magnifications. With the microscope set for a magnification of ×400 (×10 eyepiece, ×40 objective), you will see that there are about 60 erythrocytes across the diameter of the field of view. Therefore the field is about 450 µm across. At ×100 magnification, the field of view is about four times this wide – about 2,000 µm across.

PART 1 EXAMINATION OF PREPARED SLIDES

You are provided with six prepared microscope slides of bone and of a tooth. With the magnification set at ×100, examine each slide in turn:

1. *Mammal bone, transverse section (T.S.; see Fig. 5.2A).* Notice that each osteon comprises a series of concentric rings of osteocytes, surrounding the central Haversian canal. Each osteocyte lies in its own lacuna. The lacunae are somewhat larger than human erythrocytes.
2. *Human long bone (T.S.; Fig. 5.2F).* (Archaeological material, from Iran, 2000 B.C.) Notice the clearly distinguishable concentric rings of lacunae. Each lacuna is about 15 µm across. Place a piece of polarizing film over the light source and another on top of the slide. Revolve the lower film so that you cut down the light (i.e., the polaroids are almost crossed). Notice the light and dark concentric rings of each osteon (Fig. 5.2G). These are type A (alternating) osteons.
3. *Human long bone, longitudinal section (L.S.; Fig. 5.2B).* Notice that the lacunae are elongated parallel to the long axis of bone; this is the largest dimension of these melon seed–shaped spaces.
4. *Ornithomimus (ostrich-mimic dinosaur) (T.S. phalanx; Fig. 5.2C).* Notice the distinction between compact bone (light-colored) and the obviously porous cancellous bone (dark). Although you can discern Haversian canals in the compact bone, it is difficult to distinguish individual osteons. However, when viewed between crossed polarizers, the osteons can be distinguished (Fig. 5.2D). This is because the cement zone that surrounds each secondary osteon can now be seen as a thin white ring.

5. *Chasmosaurus belli (horned dinosaur) rib (T.S.; Fig. 5.2E).* Notice the osteons and the interstitial lamellae that lie between them.

6. *Hadrosaur tooth (L.S.; Fig. 5.2H).* In contrast to bone, the matrix of tooth material is traversed by numerous and very fine tubules. Notice that the enamel is without any tubules. Tooth material contains a higher proportion of calcium hydroxyapatite than bone – the enamel is almost entirely of calcium hydroxyapatite, hence its hard-wearing quality.

PART 2 PREPARATION OF THIN SECTIONS OF BONE

You are provided with two slabs of bone:

1. Horse cannon bone (L.S. are oblong, T.S. with obvious curvature).
2. *Allosaurus* femur (T.S. only).

You are also provided with two squares of thick glass (about 30×30 cm), two grades of carborundum (#320, which is coarse, and #600, which is fine), and some five-minute epoxy cement.

These samples have been cut and need to be glued to the microscope slide with five-minute epoxy cement. Since both samples are of compact bone, it is not necessary to impregnate them with cement. This would have been necessary if the sample had been of cancellous bone because that tends to pluck out during the grinding process if it is not given the mechanical support of cement.

Each slab of bone has a smooth side and a rough side. Start by mixing a small amount of epoxy cement, and leave it to stand for one minute to get rid of the air bubbles. Transfer a blob of cement onto the slide, place the bone sample, smooth side down, onto the slide, and press down firmly to seat it. Leave for ten minutes to set. Now you are ready to grind it down.

To do the grinding, wet one of the plate glass squares with water and sprinkle lightly with #320 carborundum powder (the coarser of the two). With a circular motion, and taking care to maintain an even pressure and keep the specimen flat, grind away. Wash under the tap every so often to see how thin the section has become. Continue grinding until the bone wafer is just thin enough to transmit light. You will be able to see some structure when you hold it up to the light.

Wash off the #320 powder. Repeat grinding with a fresh piece of glass sprinkled with #600 carborundum. This step is to smooth off the ground face and take the specimen to the final thickness. Continue grinding, checking periodically, until the specimen is thin enough for you to see the bone structure under the microscope. (Wash it off and dab dry before placing it under the microscope.)

Structures and Loads

A vertebrate skeleton, like many other engineering structures, is made up of columns and beams. We will be looking at how skeletons are constructed in the next chapter; the subject of the present chapter is how individual bones are loaded and how they are adapted to deal with these loads. This discussion will involve some further thoughts on columns and beams and a consideration of the construction of bones.

More on Columns and Beams

When columns are used in the usual manner, as vertical load-bearing structures, they are loaded in compression. Suppose an electronic stress meter existed that could be inserted into the interior of structures to measure local stresses. If such a device were inserted into a load-bearing column, the same amount of compressive stress would be detected at any particular level, regardless of whether the probe was at the surface or deep in the interior. This situation, of course, would apply only if the column were loaded symmetrically (Fig. 6.1A). If the vertical load were displaced from the center of the column, there would be a stress gradient across its width (Fig. 6.1B). Suppose the column were marked up into three equally spaced vertical sections, as shown in the figure. Provided the load were not displaced beyond the center section, the material would remain in compression throughout. This is shown by the graph beneath the column, which depicts the compressive stress as lying above the zero line. However, if the load were displaced into one of the outer sections, the graph would dip below the zero line (Fig. 6.1C). This means that the material in this part of the column is put into negative compression, that is, into tension. Columns are usually constructed from ceramics, such as stone or concrete, which are strong in compression but weak in tension. Therefore, if columns are loaded asymmetrically, and if the loads are sufficiently large to exceed the tensile strength of the material, the column will crack on the tensile side. Catastrophic tensile failure of asymmetrically loaded masonry columns was one of the main causes for the collapse of cathedrals in the past. Although I depicted compression as lying above the zero line in Figure 6.1, I must emphasize that this was only for convenience and that the convention in engineering is to consider compression as negative and tension as positive.

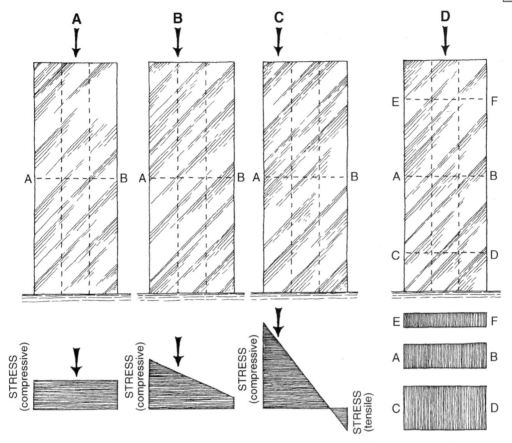

Figure 6.1. If a column were loaded symmetrically (A), as indicated by the centrally placed arrow, the compressive stress at any particular level, as at A–B, would be the same, as indicated by the shaded graph below the column. If the column were loaded asymmetrically (B), the compressive stress would be higher on one side than the other, resulting in a stress gradient across the width of the column. If the load were displaced even further toward one side of the column (C), the stress gradient would become so steep that one side of the column would be loaded in compression, the other in tension. The compressive stresses in a column increase from top to bottom (D), due to the mass of the column itself, as depicted in the three shaded graphs below the column.

The compressive stresses in a column increase from the top to the bottom, due to the mass of the column itself (Fig. 6.1D). The same is also true for tall buildings. The upper levels have to support the mass of only the top part of the building, but the lower levels have to support most of the mass of the entire building. One way to compensate for this situation would be to make the supporting columns of tall buildings increasingly thicker toward the lower levels. However, because the compressive strengths of concrete and stone are so high compared with the loads they have to bear, this is not necessary, except in very tall buildings.

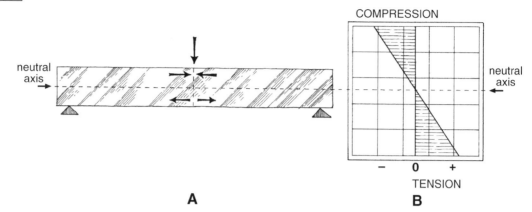

Figure 6.2. When a beam is loaded, the top edge is in compression while the bottom edge is in tension (A). This is shown graphically (B), where according to convention, compression is shown as a negative stress and tension as a positive one.

As mentioned in Chapter 2, when a beam is loaded, the top edge (actually to top surface) is in compression and the bottom one is in tension (Fig. 6.2A). Compressive stresses, as already mentioned, are said to be negative, and tensile forces are positive. If we took our imaginary probe and measured the stress in the material of the beam, we would find that the compressive stress was maximal at the top edge. If this stress were depicted on a graph, it would be shown as a maximum negative value (Fig. 6.2B). As the probe moved toward the middle of the beam, the compressive stress would decrease. Similarly, if the probe were placed on the bottom edge, it would register a maximum positive value, but this value would decrease as the probe moved toward the center of the beam. Since the top edge registers negative stress (compression) and the lower edge shows a positive one (tension), there is a level in the middle that registers no stress at all. This point is referred to as the *neutral axis*. You can visualize the neutral axis by using a beam made from a photoelastic plastic.

Photoelasticity

When certain clear plastics are mechanically stressed, they change the passage of the light rays passing through them. You can observe this phenomenon by placing a piece of polarizing material on either side of the stressed plastic, whereupon a series of colored contour lines appear. These lines, which link regions where the principal stresses are in the same direction, are more crowded together in regions of highest stress. You can then make use of this phenomenon, termed *photoelasticity*, to analyze stress patterns by constructing two-dimensional plastic models and seeing how the patterns change when the loading regime is altered. Plexiglas is sometimes cited as being photoelastic, and it is to a small degree, but there are much better plastics to use. I

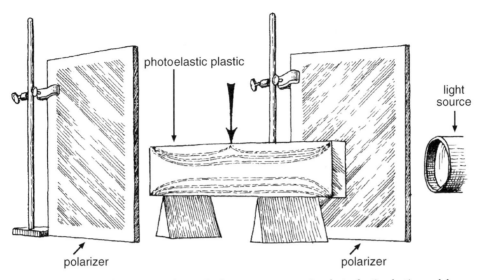

Figure 6.3. A simple apparatus for analyzing stress patterns in photoelastic plastic models.

found a class of plastic described as polycarbonate to be the best; it is manu-factured in North America under the trade name of Tuffak. If a beam of this material is set up as shown in Figure 6.3 and loaded by pressing down at the midpoint of the top edge, a series of stress lines will appear. These lines are most closely crowded together at the top and bottom edges, becoming more widely spaced toward the center. There are none in the middle, corresponding with the neutral axis.

Stresses in Beams

The implication of the neutral axis is that there are no stresses at the cen-ter of a beam loaded in bending, so there is no need to have any material there. Conversely, the stresses are maximal at the edges, so this is where most of the material of the beam should be concentrated. For beams that are sub-jected to bending in only one direction, as in a beam supporting a roof, the optimum design is the engineer's I-beam. Here most of the material is con-centrated in two horizontal bars of steel, top and bottom, called flanges, which are joined together by a vertical web of steel (Fig. 6.4A). If a beam has to support loads in two directions that are at right angles to one another, a box beam can be used (Fig. 6.4B); for multidirectional loading, a tube is best (Fig. 6.4C). Tubes are used in a wide range of applications, from bicycle frames and metal lampposts to scaffolding and the internal supports of dinosaur skeletons in a museum. The long bones of most vertebrates have a tubular construction, too. It should be noted that tubes are also stiffer, as well as being stronger, than rods of the same mass.

The stress at any point in the material of a beam is directly proportional to

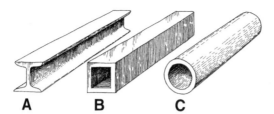

Figure 6.4. An I-beam (A), box beam (B), and tube (C) used for mechanical support under different loading regimes.

A   B   C

its distance from the neutral axis. The stress also varies with the cross-sectional shape of the beam. The cross-sectional shape of a beam is taken into account by an expression called the *second moment of area* (sometimes incorrectly referred to as the "second moment of inertia"). The equation for the second moment of area (I) of a rectangular beam is given by $I = wd^3/12$, where w = width and d = depth. The equation for a thin-walled tube of radius r and thickness t is $I = \pi r^3 t$.

Why Tubes Are Stronger Than Rods

I have a convincing demonstration of the superior strength of tubes over solid rods of the same unit mass. Using simple geometry, I calculated the dimensions for a rod and tube that have the same cross-sectional area, hence the same mass per unit length. I made suitable molds for casting the rod and tube in plaster, such that the same amount of plaster was used in each one. Once the test pieces were thoroughly dry (they were left for several days), I loaded them to failure as beams, using a simple apparatus, starting with the solid rod (Fig. 6.5). The rod always breaks at a lower loading than the tube, with a satisfying crack and clatter of falling weights.

Having a tube with the largest possible diameter places the material as far away as possible from the neutral axis, where it is most needed to resist the bending loads. However, as the diameter increases, the walls of the tube become increasingly thinner, meaning there is progressively less material to take the load. Under these conditions of thinning walls, many materials, including fibrous composites like wood and bone, become vulnerable to buckling. Buckling is where folds appear in the material being compressed,

Figure 6.5. A simple apparatus used for loading rods and tubes to failure.

causing it to bend and then to fail. The vulnerability of thin sheets of fibrous material to buckle can be demonstrated with a sheet of paper. Lay the sheet of paper on the table and load it in compression by pushing the two opposite edges closer together. Folds appear readily, and the sheet buckles.

Buckling

As tubes become progressively thinner, a point is reached when the advantage of having a large diameter to decrease the stresses acting at the surface is outweighed by the tendency for the tube to fail through buckling. There is accordingly an optimum wall thickness that gives a tube its maximum strength under a given loading regime. Buckling may be local, as when only a portion of the tube becomes rippled, but this effect, in turn, causes the whole tube to fail. You can simulate this sort of buckling with a simple party trick using an empty beer or pop can. Check to make sure the can has no dents, then carefully balance yourself on it, on one foot, steadying yourself against a suitable support. Assuming you are not exceptionally heavy, you will find that the can will support your weight, even though its walls are very thin. If you now give a small tap to the middle of the can with the edge of your other shoe, or, alternatively, get someone to give it a sharp tap with a ruler or metal coat hanger, the column will collapse catastrophically. Tapping the can puts a small dent in it, which causes more ripples to appear so that the can buckles and fails. If you were heavy enough, you would cause the can to collapse by this localized buckling without having to give it a tap first. This sort of buckling of a short, thin-walled tube loaded in axial compression is described as *Brazier buckling*. A common way of preventing Brazier buckling is to reinforce the skin of a thin-walled structure with longitudinal or circumferential thickenings, or add-ons. The fuselage of an airliner is so reinforced with longitudinal stringers and circular ribs. The raised nodes placed at intervals along the stem of bamboo and certain other grasses has a similar role. Another way of preventing buckling would be to have the structure filled with fluid. If you repeat the pop can experiment with an unopened can, it will not fail because the pressure of the contents prevents the walls of the can from collapsing. However, if you examine the can afterward, you will find that your single tap has caused a series of ripples to appear. If you drink the contents of the can and then try to stand on it, you will find that it immediately buckles and collapses.

Long thin tubes (and rods and beams) tend to fail in buckling by their entire lengths becoming arched. You can demonstrate this phenomenon by holding a vertical drinking straw against a tabletop with a fingertip and pressing down. The straw bows out to one side and collapses. This sort of failure is described as Euler (pronounced "oiler") buckling, named after one Leonhard Euler (1707–1783), a mathematician who studied the problem. Tubular structures can fail by Euler buckling whether loaded as beams or as columns. Euler

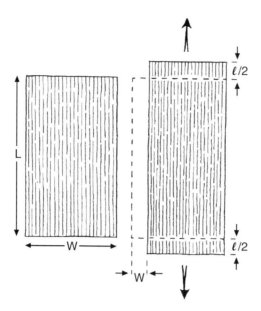

Figure 6.6. Stretching a rectangle of rubber causes elongation and narrowing.

derived an equation for the load, S, at which a column, or beam, will fail by Euler buckling:

$$S = \pi^2 \frac{EI}{L^2}$$

where E = Young's modulus of material, L = length of the column, and I = the second moment of area of the cross section of the column. For a thin-walled tube, recall that $I = \pi r^3 t$, where r = radius of curvature and t = thickness. Therefore, $S = \pi^3 Er^3 t/L^2$. If S is sufficiently high, Euler buckling can be avoided. Raising S can be achieved most effectively by increasing the radius of the tube, as this term is cubed. Decreasing the length of the tube is the next most effective strategy since this term is squared, whereas increasing its thickness is the least effective. If increasing the Young's modulus of the material is an option, that is as effective as increasing the thickness of the tube.

Poisson's Ratio

If you pull on a rubber band it stretches, thereby elongating in the direction of the tensile stress. In addition to elongating, the band also becomes narrower (Fig. 6.6). Similarly, if you took an eraser and pushed down on it, the rubber would shorten under the compressive stress and also become wider. The strain in the direction of the main stress may be referred to as the *primary strain*. The lesser strain, at right angles to the primary strain, may be referred to as the *secondary strain* (Gordon 1978). Experiments performed on a wide range of materials have shown that for each material, the ratio of the secondary strain to the primary strain is constant; this ratio is referred to as *Poisson's ratio,* after S. D. Poisson (1781–1840), who derived the relationship. As it is a ratio of strains, it has no units. Since one strain is tensile and the other compressive, they have opposite polarities, so Poisson's ratio should always be negative. However, the convention is to record the values as being positive. In Figure 6.6, the primary strain is given by l/L and the secondary strain is w/W, hence Poisson's ratio is $(w/W)/(l/L)$.

Poisson's ratio ranges from about 0.25 to 0.33 for engineering materials like metals, masonry, and concrete, but it tends to be higher for some biological materials. The values given in the literature for bone range from 0.21 to 0.62 (reported in Currey 1984).

How Bones Are Loaded

From the perspective of bearing the weight of the animal, it would make most sense if limb bones were held vertically and loaded in compression as simple columns. This is essentially what happens in elephants, which are the largest living land animals, and also in sauropod dinosaurs, which were the largest animals ever to walk on land (Fig. 6.7A, B). In both cases the long bones – the humerus, radius, ulna, femur, tibia, and fibula – are held essentially vertical. You can see this orientation when viewing them from the side or from the front. Since the bones are loaded as columns, these heavy animals maximize the weight-supporting role of their limb bones by their being solid rather than hollow. Such skeletal adaptions – columnar limb bones with solid cross sections – are referred to as being *graviportal* (Latin *gravis,* heavy; *porto,* to carry). Many other terrestrial vertebrates, in contrast, are adapted for running and are described as *cursorial* (Latin *cursor,* a runner). Cursorial adaptions, as we will see later (Chapters 7 and 10) include having the upper limb bones (humerus and femur) angled when viewed from the side (Fig. 6.7C). Notice, however, that the bones are vertical when viewed from the front or back; this is referred to as having an *erect* posture. These bones are therefore loaded as beams, albeit inclined ones, and subjected to bending stresses, which tend to load their dorsal surfaces in compression and their ventral surfaces in tension. The bones are consequently tubular in construction, the bony shafts being made of relatively thin-walled compact bone. The shafts are therefore adapted to serve as beams with multidirectional loading. Some of the forces are attributed to gravity, others to the tension generated by the muscles attaching to the bones. The individual bones, and specific regions of the bones, may be loaded in compression, tension, bending, shear, or torsion, or a combination of these.

The loading regime at the ends of the bones is different from that of the shafts because they have to transmit stresses from their joint surfaces to the shaft. This job is accomplished by cancellous bone.

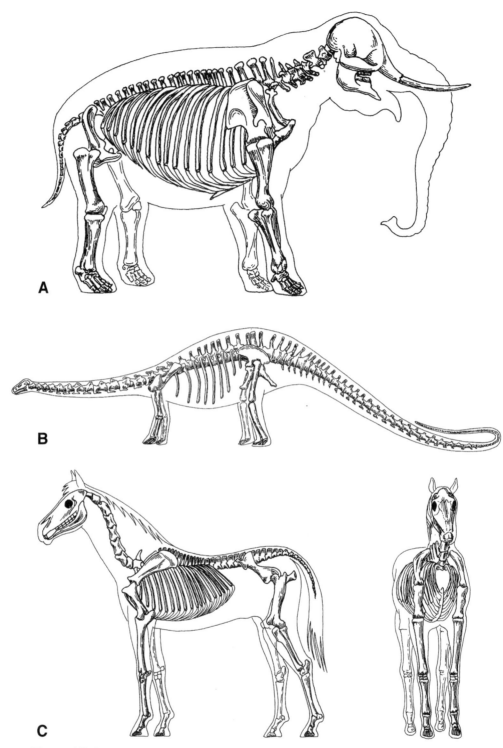

Figure 6.7. In graviportal animals, like elephants (A) and sauropod dinosaurs (B), the limb bones are essentially loaded as vertical columns. In cursorial animals, like the horse (C), the limb bones are loaded as inclined beams rather than as vertical columns. This is particularly obvious for the humerus and femur.

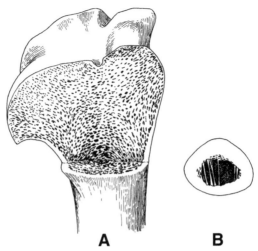

Figure 6.8. The proximal end of the humerus of a bison with a section of bone removed to show the underlying organization of the trabeculae (A). The trabeculae in the shaft (B) have the form of thin branching threads, like part of a cobweb.

A   **B**

Cancellous Bone

Cancellous bone differs from compact bone in being porous to the unaided eye, and it may indeed be visualized as compact bone with spaces. There is considerable variation in form in cancellous bone from bone to bone and from one region of a bone to another. I have the proximal end of the humerus of a bison in front of me. The shaft has been cut clean through, and a quadrant has been sliced out of the articular end (Fig. 6.8A). If I look up into the shaft I can see some fine struts of cancellous bone running across from one inside edge to the other (Fig. 6.8B). These struts are called *trabeculae* (meaning little beams). As I move the shaft from side to side I can see that these trabeculae look more like cobwebs than struts because they branch and have cross braces, too.

A close look at the central region of the articular end of the bone shows that the trabeculae have the form of thin sheets of bone. I might be convinced that the trabeculae have a random organization, like the honeycomb center of a "Crunchie" chocolate bar, or "Malteser" candy, but if I take a more distant view, I can see that there is a definite bias in their orientation. The trabeculae radiate out from the thin layer of compact bone that forms the articular surface of the bone, something like the spokes of a wheel, and converge on the sides of the shaft.

The obvious parallel between the organized arrangement of trabeculae at the ends of bones and the way that struts and braces are used in engineering structures has been recognized for well over a century (Wyman 1857, Meyer 1867, Wolff 1869). Comparisons were made between the orientation of the trabeculae and the distribution of stress trajectories in a loaded beam. *Stress trajectories* are lines of similar stress and are comparable to the contour lines on a map. Stress trajectories may be tensile or compressive, and they often intersect one another at right angles. According to the *trajectorial theory,* also

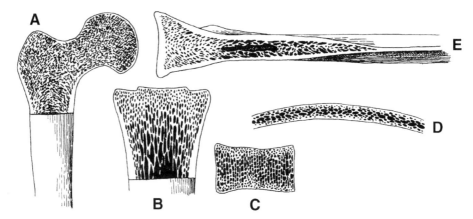

Figure 6.9. Windows cut from selected bones to show the underlying trabecular arrangement: (A) the head of a human femur; (B) the cannon bone of a horse; (C) a human lumbar vertebra (oriented as in life); (D) the cranium of a human skull; (E) the scapula of a bison.

known as Wolff's law of bone transformation, the trabeculae of cancellous bone lie along trajectories, and the density of bone reflects the intensity of the stresses. This explains why the shaft of a long bone, where the trajectories are most closely spaced, is formed of compact bone. The theory also states that if the stress patterns change, as when the loading regime is altered, the trabecular orientation changes so as to optimize load bearing. The validity of the trajectorial theory has been widely debated, and although current researchers recognize it as an oversimplification, it serves as a framework for most investigations on the biomechanics of bone. For a good discussion of the extent to which the internal architecture of bone can be correlated with the stresses acting on a bone, the reader is directed to Thomason 1995.

The relationship between trabecular orientation and the way that bones are loaded has been substantiated experimentally (Lanyon 1974, Thomason 1985). In some situations, as in the often illustrated section cut through the head of the human femur, the trabecular organization is quite complex, reflecting the multidirectional loading regime (Fig. 6.9A). In other situations, such as the proximal end of the cannon bone (third metacarpal) of the horse, which is essentially loaded in unidirectional compression, the trabecular arrangement is quite simple (Fig. 6.9B). Cancellous bone occurs in other parts of the skeleton besides the ends of long bones, but before turning to these other sites I want to say something about the mechanical properties of trabecular bone with respect to joint surfaces.

The Role of Trabecular Bone at Articular Surfaces

As we will see in Chapter 8, articular surfaces are coated with a thin layer of hyaline cartilage, whose smoothness and lubrication reduces friction. However, the strength of cartilage is considerably less than that of bone, so it cannot withstand the same large stresses. Consequently, the ends of long bones are inflated to increase the area of the articular surfaces, thereby decreasing the stresses acting on the cartilage. If the ends of bones were made of solid compact bone, it would be very stiff and would absorb little strain energy. Instead, the ends are formed of trabeculae that are oriented normal to the surface. The mechanical properties of trabecular bone are a combination of structural and material attributes. This is because the individual trabeculae act as small beams and columns, as well as acting as small pieces of bone material. Since the trabeculae are free to bend slightly, a given stress will result in a somewhat larger strain, so the Young's modulus of trabecular bone is less than it is for compact bone. Having a zone of trabecular bone beneath joint surfaces would therefore appear to impart more resilience, allowing more energy to be absorbed. However, for reasons discussed elsewhere (Currey 1984), this is not the case, except where the stresses are sufficiently high to cause plastic deformation of the trabeculae.

Other Locations for Cancellous Bone

Short bones, like carpals (in the wrist), tarsals (ankle), and vertebral centra, are thin capsules of compact bone filled with cancellous bone. The trabeculae may have an obvious orientation, as in human centra, where they are arranged longitudinally, corresponding to the direction in which they are loaded in compression (Fig. 6.9C).

Cancellous bone also occurs as a sandwich between two sheets of bone, as in the bones roofing the skull and in the scapula (Fig. 6.9D and E). Here the cancellous bone functions to stiffen and keep the two sheets apart, and also to resist shear. Consequently, there is no obvious trabecular orientation, except in the vicinity of joints, as in the acetabular region of the scapula. The same construction principle is used in industry for a variety of applications. Plastic foam, for example, is sandwiched between two sheets of thin card to produce a stiff lightweight material called foam board, used for display signs. Corrugated cardboard is made by enclosing a rippled core of cardboard between two flat sheets, and a similar process is used for making inexpensive doors. The thin wings and control surfaces of missiles are often made by bonding two sheets of metal to a honeycomb core.

The areas to which muscles attach to bones are often raised as ridges or other protuberances, and these insertion areas have a core of cancellous bone. The line of action (pull) of a muscle often has a constant relationship with the bone, so the trabeculae beneath the insertion area frequently have a particular orientation.

Measuring Loads in Structures

Engineers do not measure the stresses acting in structures directly. Instead, they measure the strains caused by the stresses, using a device called a strain gauge. The stresses can then be deduced, given the Young's modulus for the material (E = stress/strain, therefore, stress = strain $\times E$). If an aircraft manufacturer wanted to measure the stresses acting on an engine pylon, this would be done by cementing strain gauges to the outside of the structure – typically to the leading and trailing edges, where the stresses and strains would be maximal, and to the sides. From the strains recorded at the two sites, the stresses experienced by the pylon when the engine is under full thrust can be deduced.

The strain gauge is an electrical device that exploits the principle that when a wire is stretched its electrical resistance changes. The changes in resistance are small and are directly proportional to the length of the stretched wire. To increase the resistance of the wire without increasing the size of the gauge, the wire is doubled back upon itself, many times, and is sandwiched between two sheets of plastic. The device is cemented to the surface of the structure under investigation such that the lines of stress in the structure are parallel to the parallel elements of the continuous wire. This way the stress acts maximally on each of the elements of the gauge. The change in resistance of the strain gauge is measured in the usual way, using a Wheatstone Bridge circuit.[1] Strain gauges are calibrated such that a given change in resistance corresponds with a particular change in strain.

If the loading situation is simple – say a vertical column being compressed with a centrally placed load – the outcome is perfectly predictable. Here it is known that the *principal strain,* which is compressive, is lined up parallel with the longitudinal axis of the column, with a lesser principal strain, which is tensile, at right angles to it. Note that there are always two principal strains for a given surface subjected to a single load. These are orthogonal (at right angles) and of opposite sign, one usually being compressive, the other tensile. Simple loading regimes like this one seldom exist in real life, whether on an engine pylon of a 747 or on the tibia of a horse. It is therefore usually not possible to attach a single strain gauge to a structure and line it up with the principal strain, because this strain can usually only be guessed at. The solution to the problem is to use a *rosette strain gauge,* which consists of three gauges in one, each individual gauge being set with its longitudinal axis at an angle (45°, 60°, or 120°) to the next one (Fig. 6.10). We will consider 45° rosettes. Suppose the structure under investigation is a column being loaded in compression. Regardless of how the rosette is positioned on the structure, one of the three gauges will lie closer to being parallel with the major principal strain than either of the other two. Therefore, this gauge will register the highest of the three strains. The adjacent gauge (at 45°) will either record a lesser compressive strain or a tensile strain, depending on the orientation of the rosette relative to the principal strains. The third gauge will therefore record either a tensile strain or a lesser compressive strain, respectively. Since it is known that the two principal strains are orthogonal and of opposite sign, and since the strain, and polarity, are known for each of the gauges, the direction and magnitude of the principal strains can be deduced by simple geometry.

The presence of shear strains makes the calculations a little more complicated, but needless to say, computer programs are available to do all of these cal-

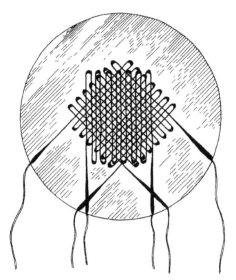

Figure 6.10. A rosette strain gauge.

culations. Thus the strains, and hence the stresses, can be read directly from a computer, given a value for Young's modulus and taking the anisotropy of the bone into account.

The loading regime of specific bones in the skeleton can be measured directly by surgically implanting rosette strain gauges onto the bones of living animals. The usual practice is to place at least four rosettes on each bone of interest: on the anterior, posterior, and on each of the lateral surfaces. By filming the experimental animal as it steps on a force plate during locomotion, you can collect external force data and use it to corroborate what was obtained internally (Biewener et al. 1983). Invasive surgery is both expensive and uncomfortable for the animal. However, Thomason and others have used a noninvasive technique for equine locomotory studies where strain gauges are cemented to the horses' hooves (see Thomason, Biewener, and Bertram 1992).

PRACTICAL

PART 1 VISUALIZING THE DISTRIBUTION OF STRESSES USING PHOTOELASTIC PLASTICS

Experiment 1   Examining stress in various beams.

You are provided with several samples of Tuffak plastic, a light source (a slide projector or other bright source), two polarizing screens (made by sandwiching a piece of polarizer between two sheets of glass), two retort stands with clamps for supporting the polarizers, one piece of frosted acetate, a piece of steel, some steel washers, a small G-clamp, and some Scotch tape (the frosted variety, not the clear type). Set up the first sample, a plain strip of the plastic, as a beam, loaded at the

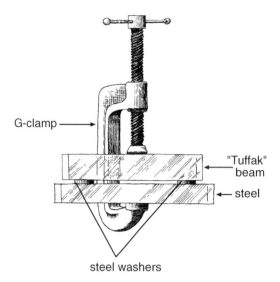

Figure 6.11. A Tuffak beam, supported by a pair of steel washers resting on a piece of steel, is loaded in the middle of its top edge.

middle of its top edge (Fig. 6.11). Make sure that the beam is vertical so that the clamp presses flat against its top edge. You do not need to apply much force to the material as it works best at fairly low stress levels. Set up the two polarizers in front of the light source, with a gap between them for the test piece (see Fig. 6.3). Rotate one of polarizers until the light viewed through the two screens appears darkest. At this point their polarizing axes are crossed. Place the piece of frosted acetate in contact with the front of the test piece (you can hold it in place with the Scotch tape). Alternatively, you can stick some of the Scotch tape to the front of the test piece and use that as the frosting (you can see the stress contours without any frosting, but it does make them much more visible).

Tighten the clamp to load the beam. Notice the symmetrical pattern of contours. Can you see how the lines are closer together at the loading point and at the support points? This pattern shows that the stress concentrations are highest in these regions. Since you are using washers as supports, you will notice areas of stress concentration at each support, with a gap in the middle corresponding to the hole in the washer. You should also be able to see that the contour lines are closer together near the top and bottom edges of the beam, with a wide space in the middle corresponding to the neutral axis. Remember that the upper portion of the beam is loaded in compression, the lower portion in tension, with no stresses in the middle.

Remove the test piece from the clamp and view it, unloaded, between the crossed polaroids. You will probably be able to see some contour lines even in the unloaded state. These lines are most likely the result of cutting the material. Cutting not only stresses the plastic but also heats it, and these patterns become "frozen in." You can remove most of them by heating the plastic with a hot-air gun after it has been cut. As a precaution, you should examine all test pieces before they are loaded so that you do not attribute any patterns that were already there to loading.

Replace the simple beam with the second sample. This one is the same as the first, except a V-shaped notch has been cut into the middle of one of its long edges. Load the beam as before, notched edge upward, with the loading point of the clamp to one side of the notch. Notice how the notch is an area of stress concentration. Turn the beam upside down and repeat. The tensile stresses on the lower edge of the beam are tending to extend the notch. Notice the stress concentration at the tip of the notch.

Replace the notched beam with the third piece of plastic, which has a hole through its middle. Notice the stress contours that appear around the hole when the beam is loaded.

Experiment 2   Investigating how stress patterns can be "frozen in."

You are provided with the ballistic apparatus used in Chapter 4 (Fig. 4.6) for breaking test samples; the apparatus used in the previous experiment; and a heat gun (a hair dryer can be used provided it has a hot setting).

Contour patterns are frozen in to the test samples by heat, as when a piece of plastic is cut with a saw. Incidentally, this is why you can see colored patterns in the windshield of a car when you look through polarizing sunglasses. The light reflected from the hood of the car is polarized, and your sunglasses, acting on the other side of the windshield, form the second polarizer. The patterns you see were frozen in by the jets of air used to cool the windshield after being formed by heating.

You can try freezing in contour patterns by loading a simple plastic beam in the G-clamp apparatus and heating it. Allow the sample to cool before releasing the stress. Once it is cool, you should be able to see some residual contour patterns permanently locked into the material.

Contour patterns can also be frozen in when a plastic is struck, which is probably due to the heat generated during the impact. Set up a sample of plastic in the ballistic apparatus, having first checked to see there were not appreciable contour patterns already there. Starting with the test piece lying flat, drop a weight onto the cutting die with sufficient force to permanently strain it. As a rough guide, a weight of 800 g at a height of 30 cm will be enough. Having bent the sample, examine it between crossed polarizers. Notice the contours around the point of impact, showing where the material was stressed.

Repeat with a larger load (1,600 g at 30 cm makes a very satisfactory dent). Notice that in addition to the contour lines around the point of impact, there are secondary areas where the test material stood on the supports of the apparatus.

Repeat with a smaller load (about 800 g at 30 cm), but this time set the sample on edge, so that the die nicks it.

PART 2 EXAMINING THE INTERNAL STRUCTURE OF BONES

Observation 1   Examine a longitudinal section through the femur of a horse (Fig. 6.12).

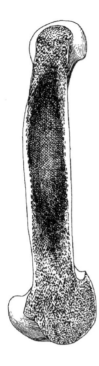

Figure 6.12. A longitudinal section cut through the femur of a horse.

This specimen is Recent (not fossil) but was buried in a field for several years. All of the marrow material – fat, blood, and soft tissue – has disappeared, leaving a dry and fairly brittle bone. Notice the outer layer of compact bone and the inner cancellous bone. The compact bone, which does not appear porous to the naked eye, forms the cylindrical shaft (diaphysis). Notice that the compact bone is thickest in the midshaft region, where bending moments are highest, and tapers toward either end. The portion of the hollow shaft that is not occupied by cancellous bone is the marrow cavity. In young animals this cavity is filled with hemopoietic tissue (erythrocyte-forming), but in mature animals it becomes filled with fat.

Notice the structure of the cancellous bone, especially the architectural arrangement of the trabeculae at the ends of the bone. They are oriented in a way as to distribute stress from the articular surface to the shaft.

Observation 2   Examine a longitudinal section through the humerus of a pelican (Fig. 6.13).

Notice the exceedingly thin wall of the diaphysis and the relatively small amount of cancellous bone, which is largely organized into discrete struts, thereby strengthening and stiffening the bone. This form of construction, paralleled in the aircraft industry, is a strategy for weight reduction, so important in flying animals. It is especially important to reduce the weight of wing bones because this reduces their inertia. The term *inertia* describes the tendency for a body to remain at rest or at constant speed, and will be discussed further in Chapter 10.

Observation 3   Examine a short section through the femur of the dinosaur *Allosaurus* (Fig. 6.14).

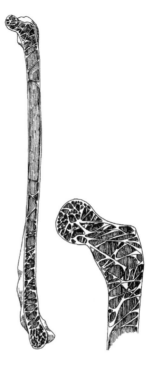

Figure 6.13. A longitudinal section cut through the humerus of a pelican.

Notice how thick the walls of the diaphysis are and how there is virtually no cancellous bone – at least in this region of the shaft. (The black color is largely due to manganese that has permeated the bone.) The marrow cavity is filled with a crystalline material (quartz). The thickness of the walls of the diaphysis suggest that the animal was relatively heavy.

Observation 4   Examine the broken portion of the radius of a mastodon (Fig. 6.15).

Notice that there is virtually no marrow cavity – the cross section of long bones of graviportal animals shows essentially solid bone.

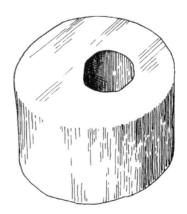

Figure 6.14. A transverse section cut through the femur of the dinosaur *Allosaurus*. Note the narrow marrow cavity.

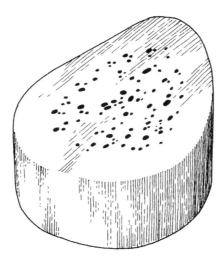

Figure 6.15. A transverse section through the humerus of a mastodon. The shaft lacks a marrow cavity.

PART 3 INVESTIGATING EULER BUCKLING

Experiment 1 Examine how the length to width ratio of a column affects Euler buckling.

Using the scissors, make a series of drinking straws of lengths 5, 10, 15, and 20 cm. Leave one straw uncut. Starting with the shortest straw, do the following for each one in turn. Measure the length and diameter of the straw. Calculate the ratio of length to diameter. While holding the straw vertically on the table with one hand, apply a gentle pressure to the free end with the tip of a finger of your other hand. A gentle push is all you need. Does the straw buckle?

You should have found that the shorter straws are resistant to buckling. However, when the ratio of the length to the width of the straw approaches about 50, it buckles very readily. Thus only a small stress is required to cause the straw to bow out at the side, which leads to the straw becoming kinked and finally failing in Euler buckling. As a very rough guide, Euler buckling occurs readily when the length to width ratio of the column or beam is about 50.

Do you think the humerus of the pelican you examined is very likely to fail by Euler buckling?

Engineering a Skeleton

The objective of this chapter is to look at some vertebrate skeletons and see how they work from an engineering perspective. I must confess that when I was a student I was unimpressed by the usual textbook comparisons between vertebrate skeletons and bridges, perhaps because neither one was explained very clearly. However, the engineering knowledge I have acquired since that time – mostly from designing experiments and teaching this course – has changed all that. I can now look at a skeleton and make some reasonable deductions about how it works, or why it has particular features, simply by interpreting the structures from an engineer's perspective. I can illustrate the usefulness of this approach with a couple of simple examples from my own research on ichthyosaurs.

The ichthyosaurs, very popular during Victorian times but no longer so fashionable amid the present obsession with dinosaurs, are a group of marine reptiles that were contemporaneous with dinosaurs. Highly modified for life in the sea, with fins instead of legs and a sharklike tail for propulsion, they were superficially similar to dolphins and sharks. One of the features of their anatomy that used to puzzle me is the way that so many of the early Jurassic species have polygonal bones in their fins (Fig. 7.1A), whereas their forebears in the Triassic have more regularly shaped finger bones (Fig. 7.1B).

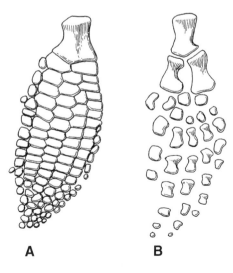

Figure 7.1. The forefin of (A) a Lower Jurassic ichthyosaur (*Ichthyosaurus communis*), and (B) a more primitive Lower Triassic one (*Grippia longirostris;* after Motani in press).

A   B

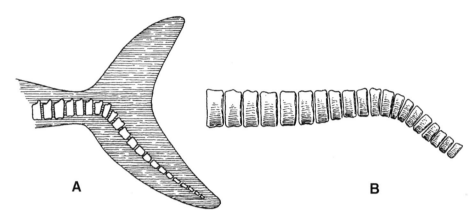

Figure 7.2. The tail, or caudal fin, of a typical ichthyosaur was supported by the down-turned vertebral column (A). Close examination of the apical vertebrae reveal their wedge-shaped profile (B).

These differences are not at all puzzling from an engineering perspective, though. The polygonal bones fit together tightly, their contact edges forming immovable joints to give a rigid fin, whereas the other fins were probably more flexible.

The ichthyosaur's crescentic tail (Fig. 7.2A) was supported by the down-turned vertebral column, a feature once thought to be a postmortem effect (Owen 1840). The reason why the vertebral column is kinked is because some of the vertebrae at the root of the tail have wedge-shaped centra (Fig. 7.2B). Some ichthyosaurs, like *Eurhinosaurus,* appear to have very steeply down-turned tails, with angles of about 50°. One paleontologist, believing this downturn to be an artifact due to the way the skeletons had been prepared, concluded that there was no tailbend in *Eurhinosaurus* at all, and no fishlike tail (Riess 1986). Actually, by measuring the dorsal and ventral widths of the wedge-shaped centra, you can calculate their angular contribution to the tail-bend with simple geometry (McGowan 1989). Unfortunately, the critical vertebrae are seldom seen in skeletons, either because of the way they are lying or because of the overlying rock. I knew from looking at other ichthyosaurs that only about six to eight wedge-shaped centra were involved in forming the tailbend, but could this be enough to contribute to a steeply downturned tail? This question prompted me to take a look at an architectural analogue. Our museum is housed in a rather nice old building, with several examples of masonry arches. One of these, which overlooks the main rotunda, spans an arc of 5 m (16 ft) in a complete semicircle. None of the twenty-one dressed limestone blocks that form the arch look markedly wedge shaped, but collectively they contribute 180° to the arc, so their individual contributions are just over 8°. I knew of three skeletons of *Eurhinosaurus* where one or two of the wedge-shaped centra can be seen. The most markedly cambered of these centra makes an angular contribution of 7° to the tailbend. So if there were eight wedge-shaped vertebrae, each contributing 7°, the tailbend could reach an angle of more than 50°. But this is all speculative; I needed to examine a complete tailbend. The chances of obtaining permission from a museum to dismantle one of their *Eurhinosaurus* skeletons to measure its wedge-shaped centra were remote. However, their shape can be revealed using a CT-scanner (computer aided tomography), and to this end I borrowed a piece of vertebral column of an unprepared specimen of *Eurhinosaurus* from a colleague in Germany. This study revealed five markedly wedged centra, with three slightly wedged ones; their combined contribution was about 40° (McGowan 1990). This is not such a steeply downturned tailbend as has been depicted in many skeletons of *Eurhinosaurus,* but it is still steeply angled compared to many other ichthyosaur species. Therefore the wedge-shaped centra, like the dressed stones in the rotunda arch, collectively have a major influence on the architecture of the tail skeleton.

To help us analyze vertebrate skeletons we need to look at certain engineering structures. We will start with the cantilever, but it will soon become apparent that the distinctions between some of the engineering structures tend to blur as one merges into another.

Cantilevers, Trusses, Girders, and Bridges

The *cantilever* is simply a beam that is supported at one end (Fig. 7.3A). A diving board is a cantilever, and so is the branch of a tree. A simple cantilever has obvious limitations in terms of how long it can be and how much of a load it can bear before sagging becomes critical. However, it can be braced by a diagonal member. This brace can be a compressive strut, which props the cantilever from below (Fig. 7.3B). Some of the lampposts outside our museum are of this construction. Alternatively, the supporting member can be tensile, bracing the cantilever from above (Fig. 7.3C). Tensile supports, called *stays,* were widely used in sailing ships, as in bracing the bowsprit.

Suppose you had a cantilever supported by a stay that was used to hoist loads. This structure could be a simple derrick comprising a pole hinged to a wall and braced by a rope, with a pulley at the free end for hoisting up loads from the street below (Fig. 7.3D). Suppose you wanted to extend the whole structure farther out from the wall. You could do so by adding a diagonal strut and a horizontal stay (Fig. 7.3E). Indeed, you could keep extending its length by adding more struts and stays (Fig. 7.3F-G). Triangular lattice structures like this are referred to as *trusses.* Even the simplest form, which is no more than a braced cantilever (Fig. 7.3D), is referred to as a truss. Trusses like these, constructed from tensile and compressive members, would obviously collapse if inverted. However, if the tensile members are exchanged for compressive ones, the truss becomes rigid and can be loaded from either direction.

Trusses come in a confusing array of shapes and sizes. One of the commonest trusses, which most of us have probably seen, is a roof truss (Fig. 7.4A). If

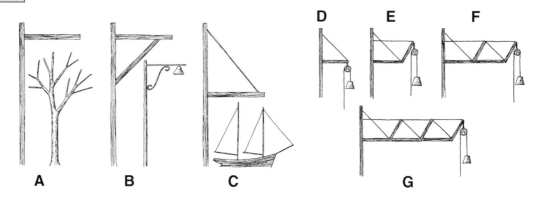

Figure 7.3. A simple cantilever (A) can be braced by a compressive strut (B) or a tensile stay (C). A braced cantilever (D) can be extended by additional struts and strays (E–G), these structures being referred to as trusses.

you have not clambered up into an attic, or seen one in a farmer's barn, you may have seen them stacked up on trucks, en route to a building site. A more complex truss, sometimes referred to as a Fink truss, was designed to bridge small gaps, as when a railway crossed a narrow ravine, and had vertical members directed downward (Fig. 7.4B). Tension in the stays prevented it from sagging under its load, but if these tensile members are replaced by compressive ones, the modified truss can be used inverted.

When large gaps have to be crossed, bridges are used, though there are some patterns of bridges that are trusses themselves. A common feature of many bridges is the incorporation of an arch. There is, for example, a bridge

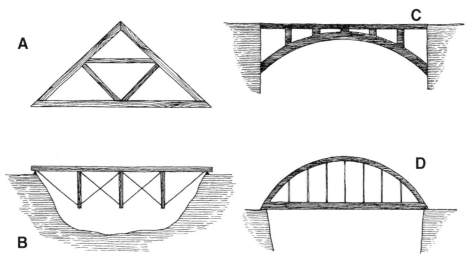

Figure 7.4. Trusses and bridges: (A) simple truss, as used to support a roof; (B) a Fink truss; (C) a simple arched bridge; (D) a tied arch, or bowstring, bridge.

over a road not far from the Royal Ontario Museum comprising a simple con-crete arch supporting a footpath (Fig. 7.4C). Arches, which are often built of steel, tend to flatten out under the load they bear, so they have to be anchored at either end by a stable foundation. That is why this type of design is often used to bridge valleys where there are rock walls at either end. An alternate design is the *tied arch* or *bowstring girder* bridge. This is an arch whose spread is prevented by a tensile member that runs, like the string in a bow, from one end to the other (Fig. 7.4D).

Now that we have an idea of some of the basic supporting structures, we can take a look at a vertebrate skeleton and see how it works.

Interpreting a Skeleton

Before attempting to interpret a skeleton, you need to bear a few points in mind. First, all we are seeing are the bare bones. Gone are the ligaments that held the bones together. Gone, too, are the muscles that also contributed to the shape of the skeleton. The missing muscles provided movement, and what a wide range of movements are available to living animals. A mounted skeleton represents a snapshot of how the bones of the skeleton may have looked at one particular instant in the animal's life. But this snapshot is an interpretation of how things might have been. Most mounted skeletons are probably reasonable approximations to reality, but we could be absolutely sure only if we took an x-ray of the animal in a particular posture, reposition-ing the bones after it had been skeletonized. As a note of caution to the liber-ties sometimes taken with skeletal mounts, I recall an experience we had with a horse skeleton. Some years ago our museum purchased the skeleton of a horse, which was to be used in a display depicting horse evolution. The new gallery was due to open within a few months, and, being behind schedule, we decided to save time and pay the supplier of the skeleton to mount it, too. We thought their familiarity with horse anatomy would have ensured a realistic mount, but we were in for a shock. Instead of depicting the horse with a char-acteristically straight back (more on this later), they gave it a strongly arched vertebral column, making it look like a giant rabbit! We had to take it all apart and mount it ourselves.

A mounted skeleton of a tetrapod, like a dog, has a number of features that would be familiar to an engineer (Fig. 7.5). The vertebral column is arched between the pectoral and pelvic girdles, the neck and tail are cantilevers (from the pectoral and pelvic regions, respectively), the long bones of the limbs are loaded as inclined beams, and the neural spines, which are especially long in the anterior thoracic region and which slope posteriorly, look as if they may form a truss. However, an engineer could not get much further than these initial obser-vations without some knowledge of the soft anatomy and the opportunity of handling the individual bones to see how they articulated. We will take a closer look at the skeleton, starting with the vertebral column. Incidentally, the dog

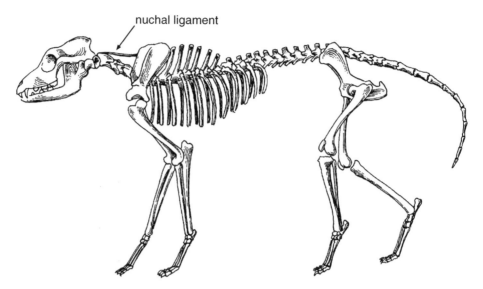

nuchal ligament

Figure 7.5. The skeleton of a dog. This particular specimen has been mounted with a rather flattened arch in the thoracic region. The only way of knowing the exact curvature would be to examine x-rays of the animal when it was alive.

skeleton illustrated here has been mounted with a rather flattened arch in the thoracic region. This depiction underscores the point made earlier that the overall shape of a mounted skeleton is in large part determined by the skills and knowledge of the technician doing the job.

The Vertebral Column

You are probably familiar with the way that different regions of the vertebral column are modified for different functions and how this reflects the specialization of the individual vertebrae. But if you are not, do not worry, because you will see it all in the practical. This regional specialization is especially marked in mammals and birds, and less so in reptiles. Mammals are also conservative in the numbers of vertebrae in each region. There are always seven cervical vertebrae, even in the giraffe with its long neck;[1] usually twelve to fourteen thoracics, which articulate with the ribs (the horse is one exception, with eighteen, and sometimes nineteen thoracics); five to seven lumbars, which have no ribs but prominent transverse processes; three to five sacrals, which are fused into the sacrum; and a variable number of caudals.

You are probably also familiar with the way that adjacent vertebrae articulate. The vertebral centra are joined together by the intervening intervertebral discs. The discs, as I will explain in a moment, are the major connections holding adjacent vertebrae together. The vertebrae are also bound together by several ligaments. The *ventral longitudinal ligament* runs along the ventral aspect of the centra and the intervertebral discs, to which it is firmly attached. It varies in development along the vertebral column. In the horse, for example, it commences in the midthoracic region as a thin narrow band, becomes thicker and wider posteriorly, and terminates by spreading out over part of the sacrum, its fibers fusing with the periosteum (the tough connective tissue that covers bones). The *dorsal longitudinal ligament* lines the floor of the neural canal and extends along the dorsal surface of the centra and intervertebral discs, to which it is fused. A cordlike ligament, the *supraspinous ligament,* attaches to the tops of the neural spines, running from the back of the skull to the sacrum. There are also the *interspinous ligaments,* sheets of connective tissue whose fibers run obliquely between adjacent neural spines. The membranous *intratransverse ligaments* connect adjacent transverse processes in the lumbar region. The vertebrae are also united by the joint capsules that enclose the articulations formed by the zygapophyses (joint capsules are described in Chapter 8).

If you held a pair of articulated vertebrae in your hands, you would see the way the articular surfaces of the zygapophyses glide past one another. Incidentally, you can always determine the correct orientation of a given vertebra by checking to see which way the articular surfaces of the zygapophyses face. The anterior ones always face upward and inward, whereas the posterior ones, like bad gamblers, are always down and out. The joints formed by the zygapophyses permit a sliding movement that allows dorsoventral and lateral movements. However, their orientation is not necessarily the same throughout the vertebral column; the changes in inclination of the articular surfaces favor lateral movements over dorsoventral movements, and vice versa. The articular surfaces of the centra do not slip past each other as the zygapophyses do. Instead, their mobility is attributed to the flexibility of the intervertebral discs.

In addition to the ligaments and the intervertebral discs that bind the vertebrae together, there are numerous small muscles running between them – the ones we injure when we have back pains. There are larger muscles, too. The muscles affect the shape of the vertebral column, sometimes contributing to locomotion, as well as helping to bind the vertebrae together. The contributions of these components in maintaining the integrity of the vertebral column were graphically demonstrated to me while I was dissecting an elephant.

Tantor, the largest elephant in captivity in North America, died at the Metropolitan Toronto Zoo on August 2, 1989, following a long surgical procedure on an infected tusk. By a long-standing arrangement with the zoo, the Royal Ontario Museum obtained the body for skeletonizing, and a small group of us spent a couple of days skinning and removing the flesh from the bones.

The animal was lying on its left side. We began by removing the right fore and hind limbs and most of the thick hide. We then decided to separate the vertebral column, just anterior to the sacrum. It took some time to cut through and remove the overlying muscle, but eventually the vertebral column was exposed and I could set about separating the ligaments and small muscles that were holding the targeted lumbar vertebrae together. To assist the separation we attached the hook of a winch through the obturator foramen, a large opening in the pelvis, and winched up the back end of the carcass, thereby placing a considerable amount of tension on the vertebral column. Wielding the sharp carving knife that had been used to remove hundreds of pounds of prime steak (unfortunately the meat could not be eaten because of all the drugs), I cut through the ligaments and muscles running between the neural spines. I thought the column would give a little, but my efforts did not seem to have any marked effect. Then I ran the knife between the articular surface of the anterior zygapophysis of the hindmost vertebra and the posterior zygapophysis of the one in front. This did make a difference, and the vertebral column yielded a little, showing that the joint capsule makes a significant contribution to holding the vertebrae together. All that was holding the column together now was the intervertebral disc, together with the ligaments and muscles, and the zygapophysial joint of the other side. Trying to find the position of the intervertebral disc was not easy, and every time I stabbed at what I thought was the correct place, I kept hitting bone. Realizing I was too far back, I tried moving forward. The tip of the knife sunk into something soft. I had found the right spot. It took very little effort to cut into the disc, and although I made a fairly small incision, the disc tore apart catastrophically. The critical crack length, which was obviously short, had been exceeded. There were two contributing factors to its short critical crack length (L_g). First, the stress in the disc was high ($L_g \propto 1/s^2$); second, the material of the disc has a fairly low Young's modulus ($L_g \propto E$; Chapter 4). The vertebral column immediately yielded, and although the ligaments and muscles of the other side, still held it together, it was very obvious that the intervertebral discs were by far the single most important structure in its integrity. Indeed, it is often reported that the vertebrae themselves are more likely to be pulled apart during traumatic auto injuries than the intervertebral discs. The elephant's discs were narrow, only about one centimeter thick, and consisted of a soft crystalline material that merged into cartilage deeper down, with white ligamentous fibers of collagen.

So much for what holds the vertebral column together, but what maintains the arched shape seen in many tetrapods? There are two primary mechanisms. First, there is an extensive ventral musculature, comprising several sheets of muscles, with their membranous tendons of attachments, of which the *external abdominal oblique* is the largest. These muscles and tendons function as tensile members – stays – to maintain the curvature of the vertebral column, the whole system serving as a bowstring girder or tied arch. Second, the vertebrae themselves are individually shaped such that they conform to the general shape of the vertebral column when they are articulated. This structure can be shown, as you will see in the practical, by attaching consecutive vertebrae together and seeing how they form a curve.

The vertebral column between the two girdles therefore functions as a compression member, with the abdominal musculature and fascia (connective tissue) in the ventral body wall forming a tensile stay. The vertebral centra are probably loaded primarily in compression, though there are tensile, bending, and shear stresses on their neural spines. Although the vertebral column is typically arched, it is quite flat in some species, such as the horse, which relates largely to differences in locomotory specializations (Chapter 10). In any event, this main portion of the vertebral column, spanning the pectoral and pelvic girdles, supports most of the weight of the animal. In this regard it is analogous to a bridge.

The cervical vertebrae, which often conform to a markedly S-shaped curve, function as a cantilever, bearing the weight of the head. Aside from the muscles that serve to raise the head, the neck is braced by an extensive and compliant ligament, the *nuchal ligament.* (Professor Michael French tells me that the nuchal ligament of the deer was used in Turkish bows.) This ligament comprises two parts: (1) a thick dorsal cord that runs from the tops of the neural spines of the anterior thoracic vertebrae and attaches to the back of the skull; (2) a pair of sheets, the laminar part, that fan out from the tops of some of the thoracic neural spines, and also from the dorsal cord, to attach to the neural spines of the cervical vertebrae. If you have a cat you can feel the cordlike portion of its nuchal ligament by feeling just in front of the shoulder blades, toward the head. If you hold it between your finger and thumb and pull on it gently, you will find that it is very compliant, like a thick rubber band. This is because the nuchal ligament contains *elastin* fibers, and elastin is a very resilient protein. The elastin is what gives the nuchal ligament its buff color; all other ligaments, and tendons, are white because of their collagen fibers. When an animal lowers its head, the nuchal ligament is stretched, and since it has such a low Young's modulus, it stores a great deal of strain energy. When the depressor muscles of the head are relaxed, the strain energy is released, raising the head with minimal muscular effort.

At this point I want to take a few moments to discuss couples and turning moments, a subject that may already be familiar to you from school physics. Imagine looking at a horse from its left side as it lowers its head to graze. The head and neck rotate, in a counterclockwise direction, about the shoulder region, largely under the force of gravity. The turning effect, or *moment,* of this movement is balanced by an equal and opposite clockwise moment provided by the tension in the nuchal ligament and neck muscles. The two equal and opposite forces balance each other and are said to form a *couple.*

Animals with large or heavy heads – herbivores like ungulates and perissodactyls immediately come to mind – typically have very tall neural spines in the anterior thoracic region. These tall spines provide an extensive anchorage for the nuchal ligament and for the muscles that elevate the head. They also provide an attachment area for some of the muscles that bind the scapula to the axial skeleton (vertebral column and ribs); the ribs provide for the rest. The pelvic girdle, in contrast, is firmly united to the axial skeleton by a bony union with the sacrum.

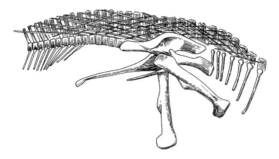

Figure 7.6. The ossified tendons in the pelvic region of a hadrosaurian dinosaur.

The relationship between the neural spines and the cross braces formed between them by the ligaments and muscles is clearly comparable to the design of certain trusses. It has also been suggested that the ribs, and their ligamentous and muscular braces, form a Fink truss (Gordon 1978, p. 239), and this idea has been confirmed by my colleague Jeff Thomason, who teaches veterinary anatomy. During his dissection classes the cadavers of horses and cows are suspended from above by wires that attach to the shoulders and to the hips. All the time the thorax is intact the animal's vertebral column is straight. However, once the ribs have been removed to expose the thorax, the vertebral column sags.

I mentioned that the tail is a cantilever, but most mammals – kangaroos being notable exceptions – do not have very extensive tails, so there is not much of a load to bear. Dinosaurs (strictly speaking, nonavian dinosaurs), in contrast, had large tails. They were important organs of balance in many groups, including all the bipedal ones, just as they are today in kangaroos. Some dinosaurs, including the elephant-sized hadrosaurs, had the front portion of their tails braced with ossified tendons (Fig. 7.6). These tendons were about as thick as a pencil and, being rigid like bone, would have made this portion of the tail stiff. The dromaeosaurs, which include *Deinonychus* and *Velociraptor* (the "raptors" of *Jurassic Park* fame), were unusual for the remarkably long anterior zygapophyses of their caudal vertebrae (Fig. 7.7). These extend so far forward

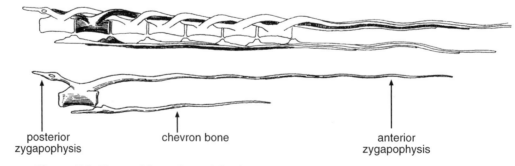

posterior
zygapophysis   chevron bone   anterior
                                                                    zygapophysis

Figure 7.7. The caudal vertebrae of the dromaeosaurian dinosaur *Deinonychus* are unusual for the great length of their anterior zygapophyses. These ensheathe the tail in a bony lattice (top).

that they overlap the preceding ten or so vertebrae, ensheathing the tail in a bony lattice. This structure made the tail stiff and ramrod straight, except at the root of the tail, which was free and mobile.

Girdles and Limbs

As noted, the pelvic girdle is firmly attached to the vertebral column by a bony union, but the pectoral girdle is strapped to the rib cage and vertebral column, largely by muscles. One could get the impression that the latter might be structurally weak compared with the pelvic girdle, especially for heavy animals like elephants. Indeed, elephants are front-end loaded, meaning that they carry most of their weight on their front rather than on their back legs. This is reflected in the large size of their front feet and is attributed to the massive weight of the elephant's head (Tantor weighed 6,500 kg, or 14,300 lb, and his head weighed 750 kg, or 1,650 lb). However, if you had seen how much muscle we had to cut away from those massive scapulae to free the pectoral girdle from the underlying skeleton, you would not doubt how securely they were attached.

The primary reason why the pectoral girdle is not immovably attached to the axial skeleton, like the pelvic girdle, is because it contributes to forelimb mobility during locomotion. This is true not only for animals like cheetahs and other cats that are so adept at bounding, but also for stiff-backed vertebrates like horses. When a horse is galloping at full speed it reaches far forward with its forelegs, and scapular mobility enhances the excursion of the foreleg, effectively adding another limb segment. Bounding animals like the big cats derive much of their locomotory power from the flexion and extension of their vertebral columns, as will be described in Chapter 10.

The limb bones are loaded as inclined beams, though in graviportal animals they are primarily loaded as columns (Chapter 6). The inclination of one bone with respect to another is maintained actively by the contraction of the muscles and passively by the tension in ligaments, at least while the animal is stationary. Although there are energy costs, there are locomotory advantages in having inclined bones, as we will see in Chapter 10. As a result of being inclined, the bones experience bending stresses, which are considerably greater than they would be if the bones were loaded as vertical columns (which is why it is easier to break a pencil in bending than in a straight pull). The bones also experience shearing and torsional stresses. Generally, those surfaces of the bones that face dorsally are loaded in compression, whereas those facing ventrally are loaded in tension. These stresses, of course, are continuously changing as the animal moves, reaching peak levels when it is running. Experiments in which animals run over force plates have shown that the peak vertical components of the ground force can be as much as about twice their body weight (Biewener 1983a), so the bending components in their bones are considerably higher.

If you have difficulty visualizing how an animal can exert twice its body weight on the ground when it is running, try the following simple experiment. Stand on a bathroom scale and wait until it settles down and records your weight. Take one foot off the scale and, while leading with this foot, take a pace forward, pushing off with your other foot. You will notice that the scale momentarily reads more than your weight. If you tried running over the scale, it would peak at much higher values. Experiments indicate that the safety factor of bones is in the range of 2 to 5, meaning that bones are about two to five times stronger than the peak stresses they experience during the height of their activity. It also seems that the peak stresses are higher in the more distal bones, so they have the smallest safety factors (Biewener 1983a). A galloping zebra is therefore unlikely to break a bone, even when fleeing at full speed from a pursuing lion. But if it does break a bone, it is more likely to be a toe bone or a cannon bone than a humerus or femur, and that is likely to be the result of a twist or a stumble, or perhaps a fatigue fracture.

PRACTICAL

PART 1 ASSESSING SHAPE AND MOBILITY IN VERTEBRAL COLUMNS

You are provided with several different vertebral columns, together with their skulls and some ribs: a dog; a horse; a duck; and the anterior part of the vertebral column of an ostrich. The individual vertebrae have been consecutively numbered. If you do not have access to any skeletons, you can still work through the descriptions by referring to the relevant illustrations.

Observation 1   How Vertebrae Articulate

Making sure you correctly distinguish between anterior and posterior, articulate any pair of adjacent vertebrae and examine the precise way they fit together. Notice the relationship between the articular surfaces of the zygapophyses. Notice also the relationship between the articular surfaces of the centra. Bear in mind that in life, these articular surfaces are invested in connective tissue. The zygapophyses are synovial joints: the movement between them is brought about by a lubricated slipping of the articular surfaces, as discussed in the next chapter.

Even though the soft tissues have been lost, we can get some idea of the degree of mobility between adjacent vertebrae by checking how much movement the articular surfaces allow. You will find that some vertebrae fit together such that a great deal of movement is permitted. Others almost lock together, indicating that there is little or no mobility.

Observation 2   The Differentiation of Different Parts of the Column

Starting with the mammals, where the vertebral counts are more conservative (seven cervicals, twelve to fourteen thoracics, five to seven lumbars, three to five sacrals, caudals variable), identify the vertebrae from the five regions (Fig. 7.8).

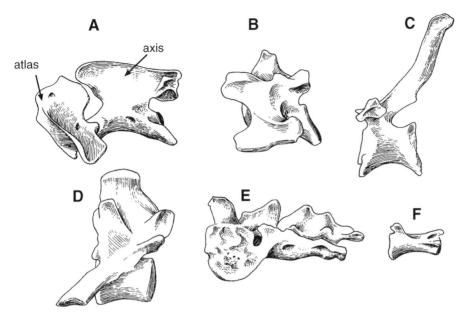

Figure 7.8. Vertebral specializations, as illustrated by the dog: (A, B) cervical; (C) thoracic; (D) lumbar; (E) sacral, and (F) caudal vertebrae.

The first two cervical vertebrae, the atlas and axis, are specialized for head support. The atlas is easily identified by the large lateral processes that extend like wings from either side. These are the cervical ribs, which are fused to the main body of the vertebra. Identify the small foramen at the base of each process. This is the *vertebrarterial canal,* for the passage of blood vessels, and represents the fork of the short cervical ribs. The primitive tetrapod condition is to have ribs that articulate with the vertebrae throughout the cervical series. Mammals are derived in having reduced cervical ribs that fuse with the vertebrae.

The atlas has no anterior zygapophyses, but there are a pair of shallow concave depressions for articulation with the paired exoccipital condyles at the back of the skull. Check to see the way the atlas articulates with the skull and how this permits a nodding movement. The axis vertebra is drawn out anteriorly into the *odontoid process* (meaning toothlike process), which locates into a corresponding *odontoid fossa* ("fossa" means depression) in the atlas. Articulate the two vertebrae and see how the axis can rotate relative to the atlas and also move from side to side. The rest of the cervical vertebrae are characterized by having vertebrarterial canals. The neural spines are short, but they become longer toward the end of the series.

The thoracic vertebrae are characterized by their modifications for articulating with the ribs, and most have long neural spines. Have a look at one of the ribs. The shorter branch of the forked end is the *tuberculum* (meaning a knob), which articulates with the short transverse process of the vertebra. The other branch is the *capitulum,* the head of the rib. The capitulum articulates with a rounded depression on the lateral aspect of the centrum, formed between the anterior edge of one and the posterior edge of the centrum in front of it. Articulate two of the thoracic vertebrae and identify the articular facet for the capitulum. Locate the capitulum of a rib into this facet, articulating the tuberculum with the transverse process.

Examine the lumbar vertebrae. They are characterized by their long transverse processes, often directed anteroventrally. Look at the sacrum and see how it attaches to the ilia of the pelvic girdle. The anterior caudal vertebrae have short neural spines and fairly prominent transverse processes, but these become reduced; the posterior ones are little more than cylinders of bone.

Observation 3   How the Shape of Individual Vertebrae Determine the Shape of the Column

1. Examine the vertebral column of the dog. When you have assessed the relative degrees of movement between adjacent vertebrae, you can try articulating them. To do so, use the wax provided to stick together adjacent articular surfaces of the centra. Use the wax sparingly. Notice the way that the shape of the vertebrae, especially the articular surfaces of the centra, determines the curvature of the whole column.

You may find some extensive overgrowths of bone on the ventral edges of many of the centra, showing where adjacent vertebrae were fused together in life. At first sight this looks like some pathological arthritic condition, but it is not. This condition is not uncommon, especially among more mature individuals, and it is called DISH (diffuse idiopathic skeletal hyperostosis). It appears that DISH develops in areas of high stress, and the fusion of the bones serves to reduce it. It has been reported in dinosaurs, especially in the caudal region of sauropods.

2. Examine the vertebral column of the horse. Because the vertebrae are fairly large, it is not very practical to try joining them using wax. Try articulating the cervicals by laying them flat along the tabletop. Does it work? You will find that it is impossible to articulate the neck flat in this way because the articular surfaces lose contact. This means that there was an arch to the neck. Repeat with the lumbar vertebrae. This works well. Horses have straight, stiff backs. Notice that the transverse processes are horizontal – in the dog they slope ventrally. Can you think of the functional significance of this difference? Think of back mobility, locomotion, and the muscles that cause flexion of the back.

3. Examine the vertebral column of the duck. Notice the extreme mobility between adjacent cervical vertebrae (there are nineteen cervicals). Look carefully at the shapes of the articular surfaces of the centra. See how these permit considerable amounts of dorsoventral, and lateral, movements. These joints are synovial, and their movements are not restricted to the compliance of the intervertebral disc. Contrast this with the rest of the vertebral column. Notice the bony splints running along the neural spines and transverse processes of the thoracic vertebrae. Birds, for reasons that will become apparent when we consider flying, have stiff backs.

4. Examine the anterior portion of the vertebral column of the ostrich. If you

could not see all the details in the previous specimen, you will be able to see them here because of the larger size. There are seventeen cervicals.

PART 2   ASSESSING MOUNTED
SKELETONS IN A MUSEUM

The purpose of this exercise is to look at some different animals in a museum and to try to relate their skeletal features with their biology. Different museums have different skeletons, and the ones discussed here were selected from those available at the Royal Ontario Museum in Toronto. I have accordingly made parenthetic notes of other species that can be substituted in cases where the same general comments still apply. If you are unable to visit a suitable museum, you can still follow the discussion by referring to the relevant illustrations.

Although the skeletons you will be examining represent animals that have long since died, try to visualize them as living animals. Think about how they supported their mass, how they moved, what particular locomotory problems they faced, and how they solved those problems. We have seen that some animals, like cats, have very flexible backs that are used during locomotion, like sprung steel. Horses, in contrast, have fairly stiff backs that play a much lesser role during locomotion. You can get some indication of the cursorial (running) abilities of an animal by comparing the length of the upper leg segment (humerus/femur) to the lower one (radius and ulna/tibia and fibula). In cursorial animals, like the horse, the lower segment is considerably longer than the upper one. In the elephant, which is not cursorial, the upper segment is longer than the lower one. Look out for this difference when you are examining the skeletons.

Locate a mastodon skeleton (or mammoth or elephant; Chapter 6, Fig. 6.7A). Identify the seven cervical vertebrae. Notice how short they are in the anterior-posterior direction. This gives the mastodon, a relative of the modern elephant, a short neck. What is the significance of the short neck? Just think about the great mass of the head and the additional turning moment due to the leverage of the tusks.[2] Notice the very large neural spines of thoracic vertebrae 2, 3, and 4. They are flared distally and are rugose (rough). Their functional significance is fairly obvious. They provided for a large attachment area for the muscles and for the ligaments involved in the support of the head. Having a short neck reduces the support problem but raises another issue. Do you think the beast could have reached down to the ground to browse? Compare its neck length with that of the Irish elk (or moose; Fig. 7.9). Notice how relatively long the vertebral centra are in the cervical region of the elk. Do you think the elk could have browsed at ground level? Now go back to the mastodon. Even if it had a long enough neck to allow the mouth to reach the ground, the tusks would have got in the way and prevented this from happening. How have living elephants resolved the problem of not being able to get their mouths close to their food? What about the trunk?

Notice the near vertical orientation of the limb bones. The individual bones

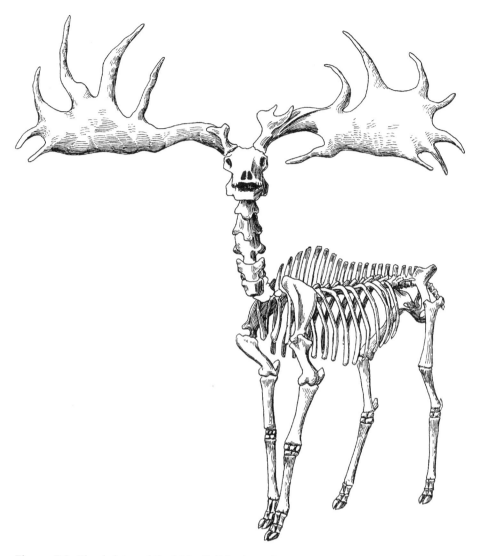

Figure 7.9. The skeleton of the Irish elk (*Megaloceros*).

are loaded as columns rather than as beams. Compare this orientation with that of the elk's. Notice in the mastodon how the acetabulum (hip socket) and glenoid (shoulder socket) are directed ventrally, so as to roof over the head of the femur and the humerus. Contrast the shoulder joint of the mastodon with your own. Consider the degree of freedom we have at the shoulder joint. The shoulder joint in the mastodon (like its hip joint) is modified for weight support, not for mobility. Notice the broad scapula. Contrast this with that of the elk. What is the functional significance? Remember that the pectoral girdle is attached to the rest of the skeleton only by muscles and ligaments – there is no bony connection.

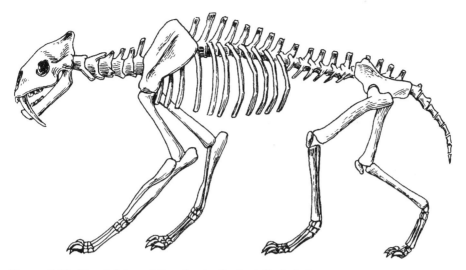

Figure 7.10. The skeleton of the saber-toothed cat (*Smilodon*).

Consider the forces tending to cause the shoulder blade to separate from the ribs in the mastodon. Having such a broad blade provided for a large attachment area. Now notice the broadly flared hips. They provide a large attachment area for the leg muscles and for the ventral muscles that attached to the ribs, functioning as stays.

Locate a modern horse (Chapter 6, Fig. 6.7C). Note the following points: straight vertebral column (no obvious arch between the girdles); seven cervicals, all relatively long, especially the axis (no. 2); long and narrow scapula. Contrast the long slender scapula with that of the saber-toothed cat, *Smilodon* (or any other cat; Fig. 7.10). In the horse, the scapula is an integral part of the running apparatus. It is very mobile and essentially functions as an additional limb segment. The vertebral column is relatively stiff. In cats, in contrast, the vertebral column is very flexible and acts as a springing mechanism as they bound along after their prey. Compare the limbs in *Equus* and *Smilodon*. Notice the difference in ratios of upper limb to lower limb. In *Equus* the upper leg (femur; humerus) is about one-third that of the rest. In *Smilodon* the upper leg is about one half that of the rest. The legs are much more muscular in the cat than in the horse. The strategy of the horse is to reduce the inertia of the leg; the cat has a muscular leg because it not only uses it for running but also for bringing down and dealing with its prey. Notice the great width of the cervical vertebrae in the cat, providing a large attachment area for the neck muscles. Cats seize their prey with their jaws, dispatching them with their powerful bite. They therefore need a powerful neck musculature, otherwise they would not be able to hang on to their struggling prey. What other comparisons can you make between the two skeletons?

Locate the *Albertosaurus* skeleton in the dinosaur gallery (or any other large

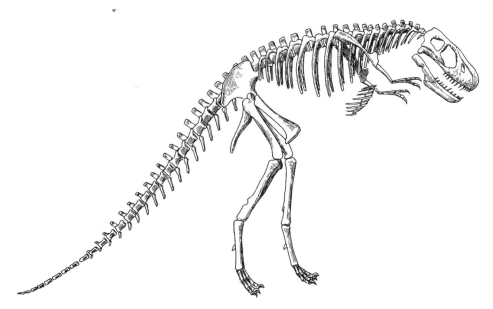

Figure 7.11. The skeleton of the dinosaur *Albertosaurus*.

carnivorous dinosaur such as *Tyrannosaurus* or *Allosaurus;* Fig. 7.11). Think about the features of the mammals you have seen, and apply the same reasoning here. Notice, for example, the long ribs that attach to the cervical vertebrae and the moderately high neural spines in the cervical region. Imagine the neck musculature of such a beast. Look at that huge head, full of sharp teeth. Contrast this situation with that of the plant-eating hadrosaur *Lambeosaurus* (or any other hadrosaur; Fig. 7.12). Here there are only very short cervical ribs and hardly any neural spines. The two animals' necks, however, are of comparable lengths.

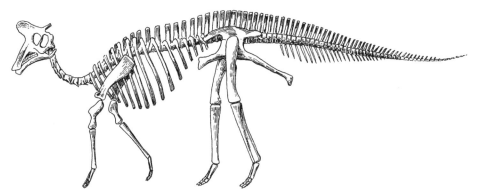

Figure 7.12. The skeleton of the hadrosaurian dinosaur *Lambeosaurus*.

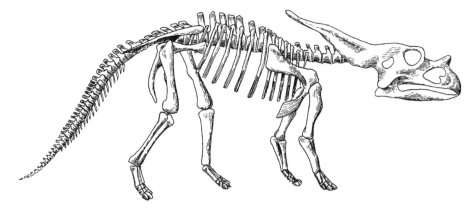

Figure 7.13. The skeleton of the ceratopsian dinosaur *Chasmosaurus*.

Locate the ceratopsian dinosaur *Chasmosaurus* (or any other horned dinosaur; Fig. 7.13). Are there any features of the skeleton that remind you of the mastodon? What about the orientation of the glenoid and acetabulum of the girdles? Compare hind limb proportions with *Albertosaurus*. Which do you think would have been the faster runner?

Friction, Lubrication, and Joints

L ike everyone else, I am always in a hurry when I walk to the bus stop in the morning. However, progress is always much slower during the winter months because of all the ice and snow. When you walk on a slippery surface you can feel the soles of your shoes slipping each time you push off. This action slows you down and uses more energy. It is because the friction between leather soles and wet snow is lower than it is when you are walking on a dry sidewalk.

Quantifying Friction

Measuring differences in friction is very simple and can be done with very basic equipment (Fig. 8.1). A block of material, with a mass W and with a smooth flat underside, rests on a smooth surface. It might be a block of wood resting on a sheet of glass. A thread, attached to one end of the block, loops over a pulley and is connected to a weight pan. A weight is put into the pan, and this weight exerts a force on the block of wood. If the negligible amount of friction in the pulley is ignored, the force, F, acting on the block is the same as the force exerted by the weights in the pan. This force tends to cause

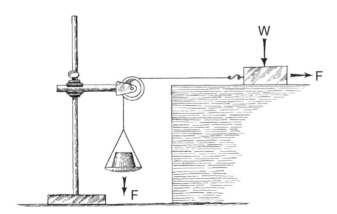

Figure 8.1. A simple apparatus for measuring the limiting frictional force.

the block of wood to slide over the glass, but this movement is resisted by the friction between the two surfaces. More weights are added to the pan until the force is just sufficient to equal the force of friction between the wood and the glass. At this point the force, F, due to the weights in the pan, is just equal to the force of friction, which is called the *limiting frictional force*.

If there were only a small amount of friction between the two surfaces, only a small amount of weight would have to be added to the pan to induce slippage. The ratio F/W would therefore be small. However, if there were a great deal of friction between them, far more weights would have to be added to the pan before the block began to slip, and the ratio F/W would be correspondingly higher. If you added extra weights to the block, thereby increasing W, the force pressing the surfaces together, you would find that a correspondingly larger number of weights would have to be added to the pan to make the block slide. Plotting a graph of F against W would give a straight line, showing the constancy of the ratio F/W. Hence F/W is constant for a given pair of surfaces; it is referred to as the *coefficient of friction*, μ. The coefficient of friction between glass and glass is 0.14, compared with 0.50 for plaster and glass, 0.016 for a brass journal bearing (explained later), and 0.0075 for a human finger joint (Barnett and Cobbold 1962). The particular coefficient just described is the *coefficient of static friction*, as it involved a stationary block that was induced to move. A variation of the experiment is to keep nudging the block as more weights are added to the pan, until a point is reached when it keeps on moving. The resulting coefficient is referred to as the *coefficient of dynamic friction* (or coefficient of kinetic friction). The coefficient of dynamic friction is lower than that of static friction. This means that if you are trying to push a heavy piece of furniture across the floor, you do not have to push quite so hard to keep it moving as you did to get it moving in the first place. Since both coefficients are ratios of force, they have no units.

Friction and Surface Areas

Surprisingly, the friction between two surfaces has nothing to do with their area of contact. You can demonstrated this fact with a simple experiment. In the friction experiment just outlined, the block can be a slab, and you can obtain the first set of results with the slab's largest surface in contact with the glass sheet. Then repeat the experiment with the slab lying on its side. The area of contact is now considerably smaller, but the results obtained are the same. This seems counterintuitive – racing cars, for example, have remarkably broad tires for improved grip on the track. Although it seems to be contradictory, the two phenomena are not the same. If you have the opportunity to inspect a racing car – they often seem to turn up in shopping malls during various promotions – take a good look at the tires. They are certainly considerably broader than the tires of a regular car, but they are perfectly smooth, without any tread. If you try picking at the rubber with a fingernail you will be surprised at how soft it is and how easily you can remove small pieces. When the tire is in use, the heat generated by the friction causes the soft rubber to melt; the resulting stickiness gives it adhesion with the track. This explains why you sometimes see drivers revving up their engines and spinning their wheels prior to the start of a race.

No matter how flat and smooth two surfaces may be, if they were viewed at superhigh magnification, they would appear like mountain ranges. The peaks, called *asperities,* one hundred or more atoms high, project above the terrain, forming the only contact with the other surface. Since the area of the peaks are infinitesimally small compared with the total surface area – perhaps one ten-thousandth for steel-steel surfaces – the friction between two surfaces is not affected by changes in their total areas of contact (Cameron 1981). The stresses at the tips of the asperities are correspondingly high, in the order of 10 GPa for steel-steel surfaces, which is sufficient to cause some opposing asperities to become temporarily welded together, then broken apart again. The friction generated between metal surfaces of machinery would cause considerable wear, leading to eventual disintegration. Such wear and eventual breakdown can be reduced, or eliminated, using lubrication.

Lubrication

There are several ways in which lubricants – usually oils – can be used to reduce friction. The first, described as *thin film* or *boundary lubrication,* involves coating the contact surfaces with a thin film of lubricant (Fig. 8.2A). The asperities of the two surfaces still come into contact, so the frictional force between them (F) is still directly proportional to the force pushing the two surfaces together (W). However, the thin film of lubricant reduces the frictional force so that the coefficient of friction (F/W) is correspondingly smaller. The lubricant also reduces wear, but not substantially because contact between asperities has not been eliminated.

The situation is completely changed if you use a thick film of lubricant that completely fills in all the spaces and keeps opposing asperities separated (Fig. 8.2B). Since there is no longer any contact between the two surfaces, the frictional force is no longer dependent on the forces pushing them together. Instead, the force of friction now depends on the viscosity of the lubricant.

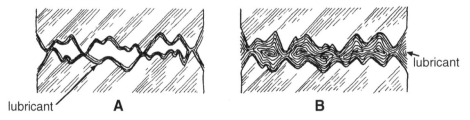

Figure 8.2. Thin film (A) and thick film lubrication (B).

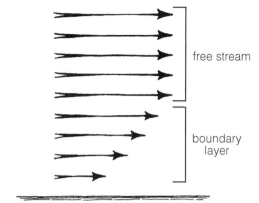

free stream

boundary
layer

Figure 8.3. Velocity gradients in a fluid
flowing over a flat surface.

Viscosity and Boundary Layers

Viscosity is a measure of how readily adjacent layers of a fluid shear or slip
past each other. Imagine a fluid flowing across a solid surface (Fig. 8.3). The
fluid molecules in contact with the solid actually adhere to the surface and are
therefore stationary. The fluid layer next to the stationary layer experiences a
friction force, tending to make it stationary, too. However, the momentum of
the rest of the fluid helps overcome the friction, causing a shearing to take
place between the two layers. The layer in contact with the stationary one
therefore moves forward, but its progress is slowed down by the friction force
between them. The same is repeated with the next layer, and with subsequent
ones, until the freely flowing stream is encountered – called the *free stream* –
where all the layers are traveling at the same velocity (the free-stream velocity).
The layers of fluid next to the surface therefore experience a velocity gradient,
and the entire zone is referred to as the *boundary layer*. If the fluid layers shear
readily, that is, if the viscosity of the fluid is low, there is more of a velocity gradient and the boundary layer is consequently thin (more in Chapter 11).

The friction between two surfaces lubricated by a thick film, as mentioned
earlier, is proportional to the viscosity of the fluid. This relationship explains
why it is harder to start a car in winter than in summer. As the temperature
falls, the viscosity of engine oil increases, thereby increasing the friction
forces between the numerous surfaces lubricated by thick films of oil.

More on Lubrication

Thick-film lubrication is an ideal solution to the problem of reducing friction and eliminating wear between two surfaces, but how can the film be
maintained? I will describe three basic mechanisms.

1. *Hydrodynamic lubrication.* The most commonly used mechanism, *hydrodynamic lubrication,* maintains a thick film by wedging the lubricant beneath
one of the moving surfaces. If an inclined plate moved across a surface coated

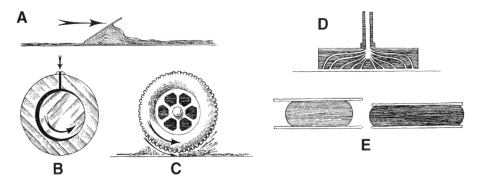

Figure 8.4. Hydrodynamic (A–C), hydrostatic (D), and squeeze film lubrication (E): (A) an inclined plate, moving across a lubricated surface, builds up a wedge of fluid; (B) a journal bearing; (C) a wheel, traveling over a wet road surface, tends to build up a wedge of water beneath it, especially when the tire is smooth; (D) high-pressure air is delivered through a series of holes to the bottom of a disc of marble; (E) increasing pressure tends to flatten a bead of fluid sandwiched between two flat surfaces.

with a layer of lubrication, the fluid would build up beneath it, and the pressure generated would keep the two surfaces apart (Fig. 8.4A). This is the underlying mechanism of hydrodynamic lubrication, and one engineering application is in the journal bearing (Fig. 8.4B). Here a circular shaft, called a journal, rotates within a surrounding sleeve, called a bush, that is filled with oil (see Cameron 1981 for an excellent account). The longitudinal axis of the journal is offset slightly from that of the sleeve, and this produces a wedging effect that squeezes oil through the gap. The pressure generated in the oil maintains the thick film, keeping the two surfaces apart.

Although most of us have probably not knowingly encountered a journal bearing, we may well have had firsthand experience with the mechanism of hydrodynamic lubrication while driving a car on a wet road. The phenomenon, usually called hydroplaning, arises when the tread on the tire cannot eliminate water as fast as it accumulates beneath the leading edge (Fig. 8.4C). A wedge of water therefore builds up in front of the tire, pressurizing the water behind it that is driven beneath the tire. The resulting thick-film lubrication separates the rubber from the tarmac, with frightening results. I recall driving along a dry highway, not so long ago, and running into a sudden cloudburst. One of the problems with front-wheel-drive cars is that the front tires wear out more rapidly than you sometimes realize. They may still look perfectly serviceable, with lots of tread, but they do not throw out as much water when driving on wet roads as they did when they were new. The sudden downpour covered the road with a thick layer of water, and my front tires, no longer able to cope, began hydroplaning. I no longer had any steering control, and when I turned the wheel to change into a slower lane, the car continued along its original course. Needless to say I keep a closer check on my tires after that scare.

The wedge angle (also called the angle of attack of the inclined plate; see

Chapter 11) can be very shallow, only 1 in 10,000, which is less than half a minute of arc. This is equivalent to an elevation of only 12 mm in the length of a football field (Cameron 1981, p. 52).

2. *Hydrostatic lubrication.* Here the lubricant is pressurized externally rather than internally as in the previous case. *Hydrostatic lubrication* therefore involves the delivery of pressurized lubricant into the gap between the two opposing surfaces, usually through a series of holes. There is a wonderful demonstration of hydrostatic lubrication at the Science Centre in Toronto. A large disc of marble sits on a marble slab, and although both surfaces are mirror smooth, the disc is so heavy that the friction force between them is very high. This makes it difficult to slide the disc over the marble surface. However, a compressed-air line, controlled by a tap, delivers high-pressure air to the underside of the disc by a series of holes (Fig. 8.4D). When the tap is opened, the compressed air is forced between the two surfaces, forming a thick lubricating layer. The disc now glides effortlessly across the marble slab, the resistance to sliding now being entirely due to the viscosity of air, which is low.

The hovercraft works on the same principle. A thick lubricating layer of air is driven beneath the craft by an engine-driven propeller. You can experience the same phenomenon next time you are boiling something in a saucepan. If you keep the lid on the pan, sufficient water condenses along the rim to form a temporary seal. If you try spinning the lid before the water begins to boil, you will find that the lid grates against the rim of the saucepan and does not move very readily. As the water in the pan begins to boil, the pressure rises in the steam and water vapor above the surface. The mounting pressure eventually forces the top of the lid to be raised, breaking the water seal and letting out a puff of water vapor. If you try spinning the lid now you will be surprised at how easily it moves, because of the thick lubricating film of water vapor separating it from the rim.

3. *Squeeze film lubrication.* This mechanism, as we will see, appears to be important in synovial joints and has engineering applications, too, as in the little ends of internal combustion engines (the joint between the piston and the connecting rod). *Squeeze film lubrication* is significant in load bearing rather than in sliding, and it is used in situations in which the duration of loading is sufficiently short and the viscosity of the fluid sufficiently high so that the thick film is not squeezed out during loading (Fig. 8.4E). Suppose, for example, that you were lowering a coverslip down onto a bead of Canada balsam on a microscope slide. The weight of the coverslip would cause the bead to flatten down, but the viscosity of the balsam is so high that it will not all be squeezed out at the sides of the coverslip. Therefore, if you wanted to slide the coverslip along a few millimeters before the balsam set, you would be able to do so because there would be a thick lubricating film of balsam beneath the coverslip. If you repeated the experiment with a drop of water, its viscosity is so much lower than that of balsam that the cover slip would squeeze most of the water out at the sides. The layer of water would therefore be so thin that the

coverslip would adhere to the glass slide by surface tension forces, making it very difficult to slide the coverslip along.

Having discussed friction between moving surfaces, and how it can be reduced by lubrication, we can now turn our attention to skeletal joints.

Classifying Joints

There is some variation in the terminology used in classifying joints, but three major types can be recognized: synarthrodial, amphiarthrodial and diarthrodial, or synovial. *Synarthrodial joints* (Greek *syn*, together; *arthron*, joint) are immovable, as in the union between the bones roofing the skull. *Amphiarthrodial joints* (Greek *amphi*, both) have a slight amount of movement, attributable to the compliance in the material uniting the bones. An example is the joint between adjacent vertebral centra, formed by the union with the intervertebral disc. *Diarthrodial joints* (Greek *dia*, throughout) are also called *synovial joints* (Greek *syn*, together; Latin *ovum*, egg), named for the synovial fluid that looks, and feels, like egg white. These joints are freely movable, like the knee and finger joints, and are what we normally think of as joints. As most of the joints of interest here are synovial, this type warrants further description.

The Synovial (Diarthrodial) Joint. The structure of the synovial joint is probably familiar to most of us through high school biology, and a standard diagram appears in most textbooks (Fig. 8.5). The diagram, which could be a knee joint, depicts a bone with a rounded distal end articulating with a fairly flat articular surface. Both articular surfaces are covered with *hyaline cartilage* (Greek *hyaleos*, glassy), which is smooth, white, and hard. It is devoid of nerves and blood vessels and is supplied with nutrients by the blood vessels in the underlying bone. Cartilage has a low Young's modulus (12 MPa) and a

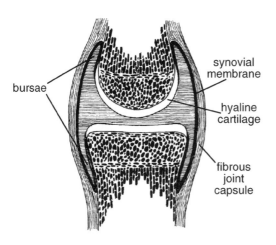

bursae

synovial membrane

hyaline cartilage

fibrous joint capsule

Figure 8.5. Diagrammatic representation of a synovial joint.

low shear modulus (4.1 MPa; values given by Torzilli 1976), compared with 6–27 GPa and 3–7 GPa, respectively, for bone (values compiled by Currey 1984). Cartilage is therefore more compliant than the underlying bone, deforming when subjected to large stresses and returning to its original shape when the stresses are reduced. It is therefore well suited to load bearing. It is also porous and permeable to fluid, which, as discussed later, appears to be related to the lubricating mechanism of the joint. Additional cartilages, called *menisci* (Greek *meniskos,* a small moon or crescent), sometimes occur within a joint between the two articular surfaces, and these are typically shaped like a segment from an orange. There may also be a mobile ring of cartilage around the articular surface, called a *marginal cartilage.* Both the menisci and marginal cartilages contain collagen fibers. This kind of cartilage is described as fibro-cartilage. Our knee joint has two prominent menisci, and their primary function is to fill in the large gaps between the articular surfaces of the femur and tibia. The reason for the large gaps is that the paired distal condyles (rounded articular processes) of the femur are markedly curved, whereas the paired depressions on the proximal end of the tibia with which they articulate are very shallow. The articular surfaces of our elbow joint, in contrast, correspond closely with one another, and this congruence eliminates the need for menisci.

The space between the articular surfaces is filled with *synovial fluid,* though it needs to be emphasized that there is barely any space at all. The human knee joint, for example, contains only about 1–2 ml of synovial fluid. This clear fluid has a viscosity that is about 1,000 times higher than that of water. The articular surfaces are enclosed by a thin membrane, the *synovial membrane,* which is well supplied by blood vessels and nerves and which secretes the synovial fluid. The synovial membrane is extended into pocket-like outpushings called *bursae.* If you have ever fallen heavily onto your knee you know that it rapidly swells; this swelling is caused by the bursae becoming filled with fluid. The condition of having an inflamed bursa is called bursitis. The synovial membrane is enclosed by the *fibrous joint capsule,* which forms a collar around the two bones. Its walls are made of interlacing bundles of collagen, the tough white connective tissue that tendons and ligaments are made of. The fibrous joint capsule is not always of uniform thickness, nor does it always form a complete collar because overlying ligaments and tendons may become incorporated within it. The ligaments that are additional to the fibrous joint capsule are called accessory ligaments. As noted earlier, synovial joints have very low coefficients of friction, with published values of between 0.002 and 0.02 (reported in Currey 1984), compared with 0.14 between two smooth glass surfaces. These numbers raise the question of how such low values are achieved.

The lubrication mechanism of the synovial joint is not well understood, but it might well involve both hydrodynamic and squeeze film lubrication. It also seems to involve a mechanism described as *weeping lubrication,* whereby fluid is squeezed out of the articular cartilage (see Torzilli 1976). This mechanism, which is neither well understood nor generally accepted, is found only in living systems. Easier to understand is how a joint, such as the knee, could function as a hydrodynamically lubricated joint. The rotation of the rounded condyles of the femur across the articular surface of the tibia could pressurize the synovial fluid, causing it to be squeezed between the two surfaces, thereby maintaining a thick lubricating film. The high viscosity of the synovial fluid also makes it likely that a squeeze film lubricating mechanism operates as well. Simple experiments with synovial fluid sandwiched between two glass plates show that it cannot be squeezed out, even at stresses exceeding those of normal loading conditions.

The types of movements made by synovial joints are largely determined by the extent of the articular surfaces and the arrangement of the surrounding ligaments. There may be gliding movements, as between the articular surfaces of the zygapophyses of the vertebrae or between the articular surfaces of the centra in mobile regions of the vertebral column like the neck. Alternatively, there may be a rolling, angular movement, as when the scroll-like distal end of the humerus tracks over the grooved articular surface of the ulna, or there may be a combination of sliding and rotating, as when the distal end of the femur rotates over the proximal end of the tibia.

Movements of limb segments are described in terms of extension and flexion. During extension, the angle between the segments increases; during flexion the angle decreases. Raising a glass flexes your elbow; reaching out your hand to shake another extends your forearm.

Regardless of how well we might be able to visualize the way synovial joints work from textbook descriptions, there is no substitute for finding out by dissection. To this end, part of this chapter's practical involves dissecting. Even if you are unable to obtain the materials used in the first part of the practical, you will surely be able to obtain a pig's trotter for dissection from the supermarket. You may even like to purchase several so that you can indulge in a great gastronomic treat afterward, namely, pigs' trotter stew, a recipe for which is given in the appendix.

PRACTICAL

In the first part of the practical you will be conducting some simple experiments to investigate solid friction. The rest of the time will be spent looking at joints. When you are looking at the specimens provided, remember that all that remains of the joints are the dry dead bones. There are no ligaments holding them together, no moist cartilage covering the articular surfaces, no joint capsule, nothing. You can get some idea of the mobility between two bones by examining their articular surfaces, but the picture is incomplete without the rest of the joint material. It is for this reason that I have chosen specimens from our own species, *Homo sapiens,* so you can check joint mobilities out for yourself.

PART 1 INVESTIGATING SOLID FRICTION

The purpose of these experiments is to investigate the force of friction – to see how it varies with the area of contact, the load force, and the nature of the surfaces. You will be able to estimate the coefficients of static friction and of dynamic (kinetic) friction from your results. The experiments are conceptually simple, but considerable care and patience are needed to achieve meaningful results. Cleanliness is of utmost importance, and the glass plate upon which your test samples are dragged has to be kept scrupulously clean – one fingerprint is enough to ruin your experiment. You are therefore provided with alcohol and tissues to keep the glass clean. You are also provided with the following: a pulley, retort stand and clamps, a set of weights (as used in the old-style chemical balances, and containing a one g, two 2 g and one each of 5, 10, 20, and 50 g weights), a set of equal-volume plaster models with towing points, some glass coverslips, thread, Scotch tape, a postcard, paper clips, and some rubber cement.

Experiment 1 Investigate the relationship between friction force and area of contact.

First make a small weight pan using the postcard, thread, and tape (Fig. 8.6). Use one of the paper clips as a suspension hook. It is important that the weight pan should hang evenly. Next, set up the weight pan, pulley, and glass plate as shown in Figure 8.1.

Taking each of the plaster models in turn (cube, cylinder, slab face down, slab edge down), loaded with a 10 g weight, determine what weights have to be placed in the weight pan to cause the model to move. Ensure the following at the start of each trial:

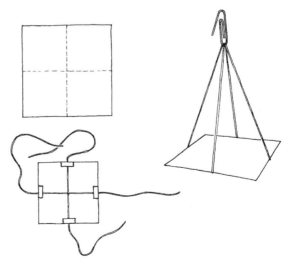

Figure 8.6. Construction of a simple weight pan.

1. The towing thread is parallel with the glass plate.
2. The model is at the same starting point (the starting line can be the edge of the book used to support the glass plate).
3. The pulley does not rub against the plate.
4. The model is stationary.

The load required to cause the model to start moving is the limiting frictional force, F (F = mass of weights in pan (kg) × 9.81 N). Very often the load will start sliding before you have attained the actual limiting frictional force. It is therefore necessary to make a number of trials to find the *maximum* load that will just start the model sliding. The procedure is to add weights (in 1 g increments when you are getting close to the sliding point) until the model just begins to slide, then to support the pan with your hand until the sliding stops. Remove your hand and see if the sliding starts again. If not, add another gram. If necessary, go back to the starting line and begin again. The reason why we use the maximum load is to achieve a reasonable level of repeatability. If you treat each trial as if you will win large cash prizes for every gram added, you will achieve a reasonable level of repeatability. Calculate the area of the surface in contact with the glass plate. For convenience, express this in square centimeters.

Repeat the experiment for the other models. Tabulate the limiting frictional force, F, in newtons, against the area of contact. Also record the loading force, W, which presses the two surfaces together. This force, of course, is the mass of the body plus the mass of the added load, in kg, multiplied by 9.81, and is in newtons.

Is there any relationship between the area of contact and the frictional force?

Experiment 2 Investigate the relationship between the loading force (W) and the limiting frictional force (F).

Repeat the previous experiment with loads of 20, 30, and 50 g added to the models. Tabulate your results. Is there any relationship between the loading force (W) and the limiting frictional force (F)?

$$\frac{\text{Limiting frictional force, } F}{\text{Loading force, } W} = \text{coefficient of static friction, } \mu$$

Calculate μ for your results from Experiments 1 and 2.

Within the limits of experimental error, you should have found that μ was constant for all your trials. Even if your results were widely variable because of experimental error, you should not have found any trends in your values. For example, ranking the shapes in terms of increasing area of contact (slab on edge, cube, cylinder, slab flat: respective areas of contact 6.2, 6.2, 7.6, and 12.5 cm^2), the values for μ with a loading force of 50 g were was 0.67, 0.55, 0.33, and 0.55, respectively. Values for μ at loadings of 10, 20, 30, and 50 g for the slab placed edge down were 0.55, 0.74, 0.75, and 0.67, respectively, compared with 0.58, 0.61, 0.60, and 0.55, respectively, when placed face down. You should have also found that the limiting frictional force was independent of the area of contact. The magnitude of μ depends on the nature of the two surfaces. The lower the value, the more easily the two surfaces slip past one another.

Experiment 3 Investigate how the coefficient of static friction varies with the nature of the surfaces.

Clean a glass coverslip with alcohol, avoiding contact with the surfaces (hold by the edges). Make a short tube from a few centimeters of Scotch tape (sticky side out) and use this to stick the glass coverslip to the bottom of a 50 g weight. Using a rubber band and a paper clip, make a towing hook and attach it firmly to the weight. Clean the glass plate with alcohol once more, making sure that it is dry.

Measure the limiting frictional force, F, as in the previous experiments, and calculate μ for glass/glass.

Coat a thin film of rubber cement on the bottom of the glass coverslip, allowing five minutes drying time. Repeat the determination of F and calculate μ for rubber/glass. If time permits, you could devise some additional experiments for measuring μ for other combinations of surfaces.

Experiment 4 Investigate the relationship between the limiting frictional force (F) and the dynamic (or kinetic) frictional force (F').

The dynamic (moving) frictional force is the force required to maintain sliding once the body has started to move. For example, take the cube and load it with a 20 g weight. Determine the limiting frictional force. By removing one gram at a time from the weight pan and nudging the model into motion, determine the dynamic frictional force.

You should have found that the dynamic frictional force is less than the limiting frictional force. Calculate the coefficient of dynamic friction, μ', where $\mu' = F'/W$. Obviously, μ' is always less than μ.

PART 2 EXAMINATION OF ANATOMICAL SPECIMENS

Observation 1 Examine the human hip (femur/acetabulum) joint (Fig. 8.7).

Is this joint from the left or right side? It is easy to tell to which side a given femur belongs because the head is inclined medially and the distal articular surface is more extensive posteriorly than anteriorly. The femur here is from the left side; so, too, is the partial pelvic girdle. Place the head of the femur into the acetabulum and move it around. This is a ball-and-socket joint and allows for a great degree of movement. The acetabulum is quite deep and cups the smooth articular surface of the femoral head. Notice how the neck of the femur is narrow. This shape allows the head to rotate within the acetabulum without its being stopped by the lip of the acetabulum. If the femur were not narrowed in this region, the degree of motion would be restricted. An alternate strategy would be to have a shallower acetabulum. The latter occurs in the shoulder (humerus/glenoid) joint. Can you think of advantages and disadvantages of the two strategies?

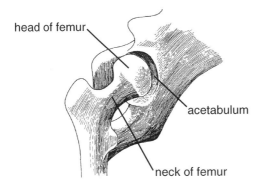

head of femur

acetabulum

neck of femur

Figure 8.7. A human hip joint.

Dislocation of the head of the humerus from the shoulder is not uncommon, but it would probably take some violence to dislocate the femur from the acetabulum. On the downside, femoral necks sometimes break, especially in older people. This probably rarely happens in the humerus.

Stand on one leg (or lie down) and make large circles with your free foot. The mobility of your leg is due to the rotation of the femoral head within the acetabulum.

Observation 2   Examine the knee (femur/tibia) joint (Fig. 8.8).

Notice the wraparound articular surface at the distal end of the femur. Articulate the two bones such that the knee joint is fully extended – that is, the leg is straight. From our own experience, we know that we cannot extend our knee joint any further than where the femur and tibia are straight in line. Keeping the tibia immobile, start flexing the knee joint by rolling the articular surface of the femur over that of the tibia. You cannot move the femur through much of an arc (about 30°) before the joint is dislocated. Obviously there has to be some slippage

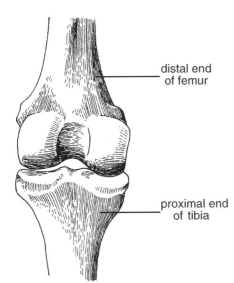

distal end
of femur

proximal end
of tibia

Figure 8.8. A human knee joint (left), shown from the posterior aspect.

of the femur upon the tibia to accommodate the high degree of flexion that the knee joint allows.

Notice that the conformation between the femoral and tibial articular surfaces is imprecise, as mentioned earlier. Indeed, there is little area of contact between them. This is best seen by viewing the joint from all angles during the flexion-extension movement. As discussed earlier, the menisci serve an important function in reducing these gaps and thereby stabilizing the knee joint.

Try rotating the femur on the tibia. This is easier when the joint is flexed than when it is fully extended. Try it out on yourself. Sit in a chair with one leg fully extended, a hand on the thigh. Try rotating your lower leg such that the toes move laterally then medially. You will find that you can only rotate your lower leg by rotating the upper leg, that is, by rotating the head of the femur in the acetabulum; if you keep your thigh stationary you cannot rotate your lower leg. Now flex your leg so that your shin is approximately at right angles to your thigh. You will now be able to rotate your lower leg without having to rotate your hip joint. To confirm that your tibia really is rotating at the knee joint, place your fingers on your shin. You will be able to feel your lower leg rotating. The reason why you cannot rotate your tibia when your leg is fully extended is due to the stabilizing action of the knee ligaments that connect the femur and tibia. There are four ligaments: the medial and lateral *collateral ligaments* (meaning parallel); and the anterior and posterior *cruciate ligaments* (meaning a cross). (See Fig. 8.9.) The medial

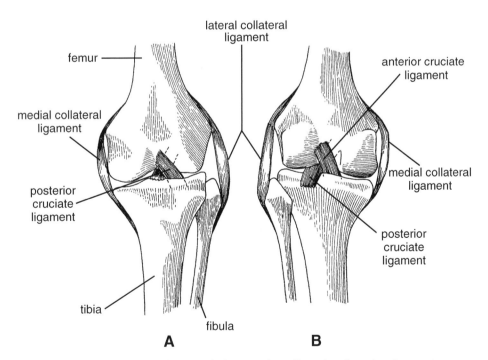

Figure 8.9. The (left) human knee joint showing the collateral and cruciate ligaments, seen in anterior (A) and posterior (B) views.

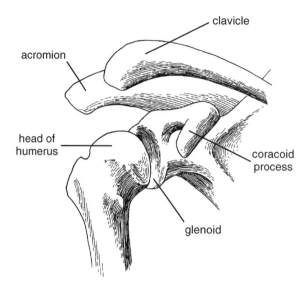

Figure 8.10. The (right) human shoulder joint, seen from the anterior and somewhat lateral aspect.

collateral ligament slopes forward, but the lateral one slopes back so that when one is slack the other tends to be taut. The cruciate ligaments form a cross and any rotational movement that slackens one tends to tighten the other.

Observation 3   Examine the articulated arm and shoulder.

Begin by looking at the shoulder joint (Fig. 8.10). In contrast to the hip joint, the socket is very shallow and the head of the humerus is not separated from the shaft by a narrow neck. This ball-and-socket joint allows for a great deal of freedom. Try making large circles with your own arm. Part of the motion you can see is attributed to the movement of the scapula upon the ribs. Nevertheless, the shoulder joint allows for a large degree of movement. When you consider how shallow the glenoid is, it is not hard to imagine the relative ease with which the shoulder can be dislocated.

Observation 4   Examine the elbow joint (Fig. 8.11).

Locate the (labeled) radius, ulna, olecranon process of the ulna, coronoid process of the ulna, coronoid fossa of the humerus, and olecranon fossa of the humerus. Flex and extend the arm. Notice that extension is limited by the contact of the olecranon process with the olecranon fossa. If there were no olecranon fossa and the posterior surface of the humerus continued without interruption, the degree of extension would be significantly reduced. Flexion is limited by the contact of the coronoid process of the ulna with the coronoid fossa of the humerus. Again, if it were not for the coronoid fossa, the extent of flexion would be much reduced.

Observation 5   Examine the joint between the proximal end of the first metacarpal and the carpus (Fig. 8.12; the first metacarpal articulates with the thumb).

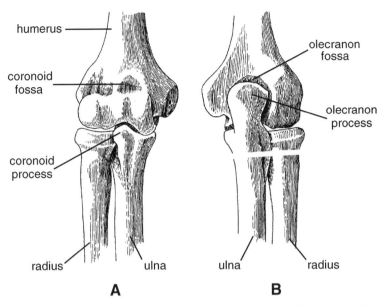

Figure 8.11. The (right) human elbow joint, in anterior (A) and posterior (B) views.

From one aspect (lateral), the articular surface of the metacarpal is convex (marked by the convex arrow) and locates into a concave articular surface on the carpal element. At right angles to this (ventral), the articular surface of the metacarpal is concave (concave arrow) and articulates with a convex surface on the carpal. Saddle-shaped articular surfaces like these allow for movements in two planes. This joint gives a great deal of flexibility to the movement of the thumb and is a feature that is not seen in other primates. Where else have you seen saddle-shaped articular surfaces?

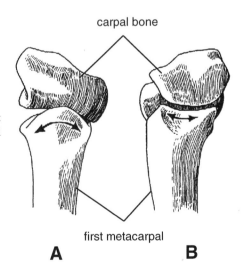

Figure 8.12. The joint between the first metacarpal and the carpus, in lateral (A) and ventral (B) views.

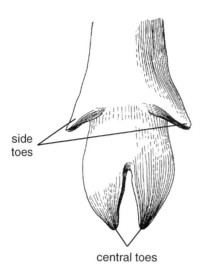

side
toes

Figure 8.13. A pig's trotter, in ventral view.

central toes

PART 3 THE ANATOMY OF A SYNOVIAL JOINT

You are provided with a pig's trotter (the distal portion of the limb, from about the level of the wrist/ankle), a syringe, and a microscope slide.

Carefully remove the skin. This is not easy; the skin is tough and is closely connected to underlying ligaments. Take care not to remove these ligaments with the skin because doing so may very well disrupt some of the joint capsules. Be patient! It will probably take you more than half an hour to remove the skin. Having removed the skin, familiarize yourself with the general anatomy of the foot, locating the two central toes, the side toes, and the metacarpus (or metatarsus in the case of the hind foot), which corresponds to the palm of our hand (Fig. 8.13). Examining the foot from one side, locate the joint between the two central toes and the metacarpus (or metatarsus) – this is a knuckle joint. By moving the toe and feeling the bones that form the joint, determine the precise region of contact between the two articulating surfaces: it may be necessary to free and clear the overlying tendons to do this. (Remember that tendons connect muscles to bones, ligaments connect bones to bones.) You now need to insert the needle of the syringe into the joint cavity to extract some synovial fluid. This is not easy to do because it is tricky to maneuver the needle into the cavity, and there is so little fluid there anyway. Proceed as follows: while holding your thumb over the dorsal aspect of the joint, flex the toe (that is, depress it). You will be able to feel your way down with the needle. Withdraw the plunger to remove a sample of fluid. At best you will collect only one or two drops, and if you are in the right place you should hear a sucking noise. If there is no indication that you have collected any fluid, move the needle a little and try again (do not withdraw the needle if you can help it – just move the tip around a bit). When you think you have collected some synovial fluid, withdraw the needle and transfer the sample to a microscope slide. Touch it and feel how slippery it is.

Fill the syringe with air, insert the needle into the hole again, and inject air into the joint cavity. You will see that extensive ballooning occurs around the joint. This is because parts of the joint capsule are thin-walled. Notice that some of the ballooning occurs at some distance from the actual joint – these extensions of the capsule, as noted earlier, are the bursae. The internal surfaces are lubricated by synovial fluid, and it is said that they facilitate the movements of the tendons and muscles that overlie the joint (Sisson 1953).

Locate one of the knuckle joints of an outer toe. Carefully cut about halfway around the joint capsule to expose the articular surfaces of the two bones. Notice that the joint capsule is essentially a thick fibrous collar that is firmly attached around the two ends of the bones. Joint capsules are generally not of uniform thickness like this: there is usually localized thickening in which the constituent fibers are arranged parallel to one another. These thickenings are the ligaments of the joint. There are often accessory ligaments, too, which lie outside the joint capsule. There are very obvious accessory ligaments on either side of the knuckle joints (take a look). Sometimes accessory ligaments lie *inside* the joint capsule, as in the hip joint. This cordlike ligament, which runs from the head of the femur to the acetabulum, serves to hold the femur securely in place within the socket. The outer layer of the joint capsule is fibrous, comprising interlacing bundles of white connective tissue. Beneath this layer lies the synovial membrane. The synovial membrane, whose cells secrete the synovial fluid, lines the entire joint capsule (and bursae), except the articular cartilages. In some joints an overlying ligament (or tendon) is incorporated into the joint capsule. Where this is the case, the internal surface of the ligament is invested in the synovial membrane. If these ligaments are removed during dissection, the joint capsule is perforated (hence the cautionary note about taking care not to remove tendons and ligaments with the skin at the start of the dissection).

In some joints, as in the knee, a fibrous pad of cartilage intervenes between the two articular surfaces. The pad may extend right across the joint, being attached all around its periphery to the joint capsule. Alternatively, the pad extends across only part of the joint, forming the menisci, referred to earlier.

Examine the partially dissected joint capsule of the side toe. Notice that the articular surfaces are white – this is due to the articular cartilage that caps the ends of the bones. Continue cutting around the joint capsule so that you can see the shapes of the articular surfaces. Flex and extend the toe to see how the distal surface (of the first phalanx) tracks over the proximal surface (of the metacarpus). The shape of the two articular surfaces partly determines the plane in which the bones move relative to one another.

Can you see a small bone (actually a double structure) on the ventral aspect, between the first phalanx and the metacarpal? Its articular surface tracks across the metacarpal surface. These additional bones are called sesamoids. The kneecap is a sesamoid bone. A sesamoid bone is a bone that forms within a tendon, and its function is to increase the moment arm of the muscle to which the tendon belongs.

Pare away some thin slices of articular cartilage. Place these onto the microscope slide and examine them under the microscope. You may recognize the histological appearance of hyaline cartilage.

Remove the side toe and have another look at the articular surface. Notice how smooth it feels.

Muscles: The Driving Force of Skeletons

For most paleontologists the joy of the hunt is the most appealing part of fieldwork, but not for me. I like smashing things – rocks mostly. Give me a crowbar and an outcrop of rock to be removed, and I am in fieldwork heaven. It is remarkable what damage can be done with a crowbar. As Archimedes said, "Give me a lever long enough, and a place to stand, and I will move the earth," or words to that effect. Unfortunately, the mechanics of levers is nowhere near as engaging as their use, but we need to be familiar with them to understand some aspects of muscle function.

Levers

Suppose you are using a screwdriver as a lever to pry open the lid of a paint can (Fig. 9.1A). You would insert the tip of the screwdriver beneath the lid and use the rim of the can as a point of purchase – the fulcrum. The force you apply to the lever is called the *effort,* and the resulting force exerted by the lever is called the *load.* The direction in which the effort force is applied is described as the *line of action* of the effort, and the same is the case for the load. In the paint can example, the lines of action of the effort and load are parallel to one another and at an angle to the lever. It might seem that the two lines of action are always parallel to one another, but this is not necessarily true, as, for example, with the jaw muscles, which will be discussed later.

The *moment arm* is the distance from the point of application of the force to the fulcrum, and it is always measured perpendicular to the line of action of the force. The *mechanical advantage* is the ratio of the force applied to the load, to that applied by the effort. If you pushed down on the screwdriver with a force (effort) of 2 newtons, and the screwdriver pushed up on the lid with a force (load) of 12 N, the mechanical advantage would be 12/2 = 6. The mechanical advantage is also the ratio of the moment arm of the effort to that of the load:

$$\text{Mechanical advantage} = \frac{\text{load force}}{\text{effort force}} = \frac{\text{moment arm of effort}}{\text{moment arm of load}}$$

Therefore, load force = effort force × mechanical advantage. Suppose I am using a crowbar to lever up a rock, and I am leaning on the bar with all of my 77.7 kg of weight (Fig. 9.1B). If the moment arm of the effort is 140 cm, and

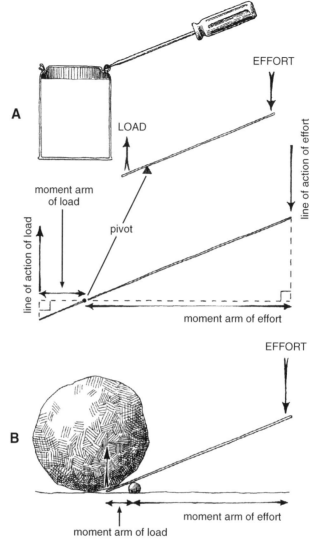

Figure 9.1. Using a lever to pry up a paint can (A) and move a rock (B). These are examples of class 1 levers, where the effort and load are on opposite sides of the fulcrum.

that of the load is 7 cm, the mechanical advantage is 140/7 = 20. The force exerted on the load is therefore 77.3 × 9.81 × 20 = 15,166 N. This is a very large force, and it might seem as if we are getting something for nothing. However, nothing in life is free, and the downside of the situation is that I have to move my end of the crowbar through a large distance just to get a small movement at the other end. The distance that I have to move my end, compared to the distance the load end moves, is numerically equal to the mechanical advantage and is referred to as the velocity ratio. The *velocity ratio* is therefore defined as the ratio of the distance the effort moves to the distance the load moves. Since the mechanical ratio is 20, if I move my end of the crowbar through a distance of 50 cm, the rock at the other end will move through a distance of only 50/20 = 2.5 cm. It follows that the work got out of the system is the same as that put in, less a little that is lost through friction. In this example, the force applied (effort) is 77.3×9.81 N, and since this force moves through a distance of 0.5 m, the work put in is $77.3 \times 9.81 \times 0.5$ joules = 379.2 J. Meanwhile, the force exerted upon the load is 15,166.3 N, and since this force moves through a distance of 0.025 m, the work got out is $15,166.3 \times 0.25$ joules = 379.2 J.

Levers have been classified according to the relationship between the effort, the load, and the fulcrum. In the examples given so far, the effort and load are on opposite sides of the fulcrum. These levers are described as *class 1 levers*. Other examples of class 1 levers include pliers and scissors (Fig. 9.2A), and the action of the triceps muscle in pulling on a rope held in the hand (Fig. 9.2B). When the effort and load act on the same side of the fulcrum, the levers are described as class 2 or class 3, depending on which of the two forces is closest to the fulcrum. When effort and load act on the same side of the fulcrum, with the effort farther from the fulcrum than the load, you have a *class 2 lever* system. The mechanical advantage of class 2 levers is greater than one, and a familiar example is a wheelbarrow (Fig. 9.2C). Class 2 levers are not common in vertebrate skeletons, but one example is the gastrocnemius (calf) muscle used when we raise ourselves onto our toes (Fig. 9.2D). A *class 3 lever* is where the effort and load act on the same side of the fulcrum and the effort is closer to the fulcrum than the load. An example from the physical world is a pair of forceps (Fig. 9.2E). Many of the skeletomuscular systems are class 3 levers, including the adductor (closing) muscles of the jaw (Fig. 9.2F) and the biceps muscle.

Whereas the mechanical advantage of our simple machines is usually greater than one, that of musculoskeletal systems is usually less than one. The forces exerted by the muscles are therefore more than those exerted on the loads, often very much more. This might appear to be an unsatisfactory situation, but it is an inevitable consequence of a muscle's limited degree of contraction. Muscles, at most, can shorten by about one-third of their length, which imposes limitations on the amount of movement they can effect. The human biceps, for example, is about 21 cm long, so the maximum distance over which it can exert an effort is only about 7 cm. However, this distance is sufficient to move the distal end of the forearm through almost 180°. If the muscle were attached far from the joint, movement would be severely restricted, outweighing the improvement in mechanical advantage. The other aspect of having a low mechanical advantage lever system is that the speed of movement of the distal end of the limb is increased. For example, in the time it takes my biceps muscle to contract, moving my fully extended forearm into a fully flexed position, the point of attachment of the muscle moves through about 7 cm. But my wrist moves through about 70 cm, meaning that it moves about ten times faster. The speed of movements of distal limb segments is of great significance in locomotion (Chapter 10).

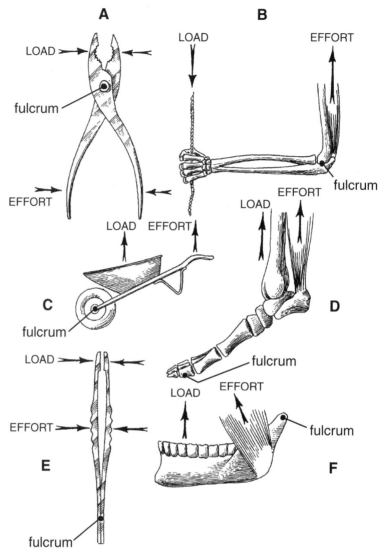

Figure 9.2. Examples of class 1 levers (A, B). Class 2 levers have effort and load on the same side as the fulcrum, with the effort farthest away (C, D). Class 3 levers have effort and load on the same side of the fulcrum, with the effort closest (E, F).

Skeletal Muscles

Before dealing with the anatomy and fine structure of skeletal muscles, I want to state the obvious: muscles can only contract. Consequently, each joint requires at least two muscles: one to extend, thereby increasing the joint angle, and the other to flex, decreasing the angle.

The simplest skeletal muscle has a *tendon of origin* at its proximal end and a *tendon of insertion* at its distal end (Fig. 9.3). The body of the muscle is

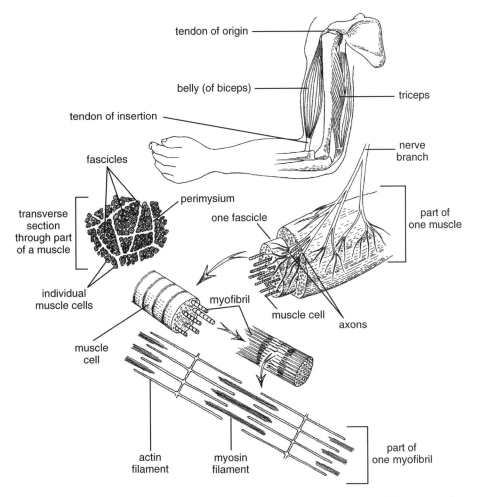

Figure 9.3. The anatomy of skeletal muscles and how they are innervated by nerves. See text for details.

referred to as the *belly*. The belly is enclosed in a tough connective tissue sheath (mostly of collagen) called the *epimysium*. The easiest way to understand the structure of a muscle is to look at transverse sections cut through one of them. I should warn you, though, that it is a bit like looking at those Russian dolls, one inside the other and all looking much the same. It is therefore sometimes easy to forget at what level you are looking.

The first structures seen beneath the epimysium are the *fascicles*. Each fascicle is a bundle of muscle cells and is enclosed in a sheath of connective tissue called the *perimysium*. It is the fascicles that give a skeletal muscle its fibrous appearance in gross structure. Indeed, the fascicles are often referred to as fibers, which is most unfortunate because so, too, are the muscle cells themselves. The only way of knowing in which sense the term muscle "fiber" is being used is to consider the meaning of the whole sentence.

Fine Structure If you look at a single fascicle under the microscope, you can see the individual muscle cells separated from each other by the *endomysium*. These individual muscle cells, or fibers, are long and narrow, about 10–40 μm in diameter, and they may run the entire length of the belly of a muscle, apparently reaching lengths of over 30 cm in some muscles in our own species (Williams and Warwick 1980). This is remarkably long for a cell, but the functional significance is that muscle cells contract by decreasing their length. Imagine what a minuscule contraction a muscle cell would effect if it were the size of a red blood cell. Muscle cells are multinucleated, and the nuclei, which are elongated in the direction of the long axis of the cell, are arranged peripherally. Transverse sections cut through fascicles therefore usually reveal at least some nuclei. Packed within each cell, like drinking straws in a box, are fine fibers called *myofibrils*, which have diameters of about 1 μm. Each myofibril is packed with *myofilaments*, of which there are two kinds, actin and myosin. The actin and myosin filaments interdigitate, giving skeletal muscle its characteristic banded appearance when viewed under a microscope and providing the contractile mechanism.

Muscle Pennation At the simplest level of organization, muscle fibers (fascicles) run parallel to the belly of the muscle, from the proximal tendon of origin to the distal tendon of insertion (Fig. 9.4A). When the muscle cells contract, shortening by one-fourth to one-third of their length, the entire muscle accordingly contracts to the same extent. An example of this type of muscle, described as being nonpennate (Latin *penna,* feather), is the sartorius, a long, straplike muscle of the thigh (the longest muscle in the human body; Williams and Warwick 1980). Far more common are *pennate muscles,* so called because the oblique arrangement of their fibers gives them a superficial resemblance to a feather. Pennate (often spelled pinnate) muscles come in a variety of forms, of which there are two main types, unipennate and bipennate. In *unipennate* muscles, the tendon of origin is an extensive sheet of connective tissue, called an *aponeurosis of origin,* that covers much of one side of the belly. Similarly, the tendon of insertion is an extensive sheet, called the *aponeurosis of insertion,* that covers much of the opposite side of the belly, and the muscle fibers (fascicles, remember) run obliquely between the two (Fig. 9.4B). In *bipennate* muscles, the aponeurosis of origin occupies much of the surface of the belly, and there is a central tendon of insertion onto which fibers converge from the origin (Fig. 9.4C). The *angle of pennation* is the angle subtended by the fibers to the aponeurosis of insertion (Fig. 9.4D). If the angle is shallow, the fibers are relatively long and few in number (Fig. 9.4E). The fibers become shorter and more numerous with increasing angle (Fig. 9.4F).

Pennation is an elegantly simple mechanism for effecting large functional differences between muscles. By increasing the angle of pennation, and by being bipennate rather than unipennate, more fascicles, and hence more muscle cells, are packed into a given volume of muscle tissue. The force of contraction is directly proportional to the sum of the cross-sectional areas of

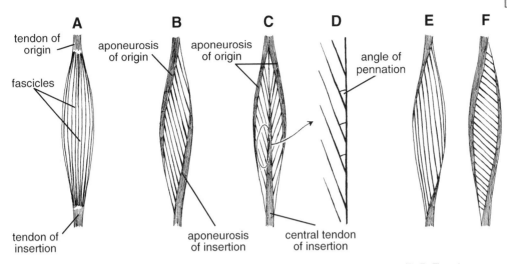

Figure 9.4. The internal architecture of skeletal muscles: (A) nonpennate; (B, E, F) unipennate; (C) bipennate.

the muscle cells. Consequently, pennate muscles generate more force than nonpennate ones of similar volume. It then follows that bipennate muscles generate even more force than unipennate ones of similar volume, and the force increases with the angle of pennation. Although pennate muscles generate greater forces by virtue of their having a larger number of muscle cells, they do not shorten very much during contraction because their cells are so short. In some situations a large excursion of the joint is the primary role of a given muscle, and it is therefore likely to be nonpennate. In other instances, the force of contraction is primary, and the muscle is accordingly pennate, and often bipennate. A familiar example of this case is the bipennate muscle we find when we crack open a lobster's claw. The central tendon of insertion is the stiff translucent structure in the middle of the muscle that the short fibers converge upon (called an apodeme). When the lobster closes its claws the amount of movement is minimal, but the force of its contraction is powerful enough to cause serious injury to any fingers that might be in the way.

Skeletal muscles therefore possess a wide range of functional diversity through their architecture. But they also have another level of functional diversity that is expressed through the physiology of the individual muscle fibers.

Muscle Physiology There are three major types of skeletal muscle cells. They differ in the speed and force of their contractions, and in their endurance. You can identify them histologically through their differential staining reactions for respiratory enzymes, but before discussing their differences, I need to say a few words about aerobic and anaerobic metabolism.

When glucose is respired it becomes broken down into pyruvate. If oxygen is available, the pyruvate enters into Krebs cycle, where it is further broken down, the end products being energy, in the form of ATP, carbon dioxide, water, and heat. This process is referred to as *aerobic metabolism,* and it takes place within the mitochondria. If oxygen is not available, Krebs cycle is not entered. Instead, the pyruvate is broken down into lactate, with the release of lesser amounts of energy, again in the form of ATP, together with carbon dioxide, water, and heat. This anaerobic process is called *glycolysis.*

The three main types of muscle cells are slow-oxidative (SO), fast-oxidative/glycolytic (FOG), and fast-glycolytic (FG). (An excellent review is given by Peters 1989.) *Slow-oxidative* fibers are often referred to as red fibers because of the color imparted by their high myoglobin content. They are rich in mitochondria, have the slowest contraction speed, and have the smallest force of contraction, but they have the highest endurance because of their high oxidative capacity. Their glycolytic capacity – that is, their ability to respire anaerobically – is low. *Fast-oxidative/glycolytic* fibers have high contraction speeds, an intermediate contraction force, and high endurance. They have moderate to high oxidative and glycolytic capacities, and like the slow-oxidative fibers, they are red. The fastest of the three are the *fast-glycolytic* fibers, often called white fibers because of their pale color. They contract with the greatest force, but they have the lowest endurance due to their low to moderate oxidative capacity.

It is common for certain muscles, or for certain parts of particular muscles, to predominate in one of the three fiber types. The pectoral (main flight) muscles of chickens and other galliform (game) birds, for example, consist largely of fast-glycolytic fibers, hence the white color of the breast meat. The pectoral muscles of ducks, in contrast, predominate in fast-oxidative/glycolytic (FOG) fibers, and are deep red (see Butler 1991 for further discussion). These differences are reflected in the marked functional differences in their respective flying performance. Game birds are good sprinters and can take off as if rocket-assisted, but they cannot keep it up for very long. Ducks, in contrast, take a lot longer to become airborne, but they have great stamina and can keep flying for hours on end.

Motor Units A given anatomical nerve – those white cords one sees during dissection – comprises a bundle of individual axons for transmitting impulses. Nerves that transmit impulses to muscles are called motor nerves, and each skeletal muscle is supplied with a motor nerve or a branch of one. Each axon has a number of nerve endings, and each one supplies one muscle cell. A given axon therefore supplies several muscle cells, and the whole unit, the axon and all the muscle cells it supplies, is called a *motor unit.* Experiments have shown that a given motor unit comprises all the same kind of muscle cells. Those made up of slow-oxidative cells are called *slow motor units,* those containing fast-oxidative/glycolytic cells are called *fast fatigue resistant,* and those comprising fast-glycolytic cells are called *fast fatigable motor units.*

The Graded Response of Muscles Muscle cells, like the nerve cells that supply them, have an all-or-nothing response. This means that either they respond to a stimulus with a maximal contraction, or they do not contract at all. There are no graded responses. But we all know through experience that our skeletal muscles have graded responses. I can move one leg slowly and cross it over the other while I am sitting down in a chair, or I can move both legs as fast as I possibly can when I am sprinting for a bus. This movement is all achieved through the number of motor units recruited into action. When we are making minimal demands upon our skeletal muscles, relatively few motor units are called into play. But as the demand increases, more motor units are involved, and when maximum power is required, all motor units are functional. Significantly, there is a graded response in the type, as well as in the number, of motor units that are activated during increasing levels of activity. At first, slow motor units are recruited, which serve for posture and for slow movements. As activity levels increase, fast fatigue resistant motor units are brought into action. Finally, the fast fatigable motor units are recruited as the skeletal muscles generate their maximum power output.

Cross-innervation experiments have shown that motor units are able to undergo transformation from one type to another. For example, if the nerve supply to a slow-oxidative muscle cell and to a fast-glycolytic cell were severed, and the respective nerves were swapped over, the slow muscle would transform into a fast one, and vice versa. Transformation of one muscle cell into another also occurs during training. For example, endurance activities, such as long-distance running, can cause fast cells in the leg muscles to be converted into slow ones (Goldspink 1991).

Exercise Physiology

Exercise physiology is a large and active field of research, with an extensive literature, and it could easily take an entire book to do justice to the subject. I can therefore give only an overview of the subject here to serve as a prelude to the last four chapters, which deal primarily with locomotion. According to the *English Oxford Dictionary,* exercise is "the exertion of muscles, limbs. . . ." This is a useful definition for our purposes because exercise is all about the exertion of muscles, and it is mostly, but not exclusively, to do with locomotion. Non-locomotory exercise activities include such things as nest-building, burrowing, and feeding. There are also many organisms, like barnacles, that are sessile (attached) but still active. When barnacles are covered by the sea, they actively propel food toward their mouths by the rhythmic movements of their feeding appendages. However, we are concerned here only with the locomotory aspects of exercise – walking, running, flying, and swimming.

Sprinting and Long-Distance Performance

Some individuals are more active than others, and the same is also true among different species. But how are activity levels compared? Before considering this, we need to make a distinction between two primary aspects of exercise, namely, burst speed, or sprinting, and endurance speed. *Burst speed* is the greatest speed an animal can attain over a short distance, or for a short time. *Endurance speed* is the greatest speed an animal can sustain over a long distance, or for a long time. What constitutes a long distance or a long time? When hunting dogs cover distances of several kilometers, or keep up their trotting in search of prey for several hours, they are displaying endurance exercise. The ability to maintain high levels of exercise over a long time is referred to as *stamina*. A cheetah, in contrast, may sprint at its top speed for less than 100 m, or for a matter of seconds, which is described as burst activity. Some animals are specialists in sprinting, whereas others are specialized for endurance. The wildebeest, for example, typically ranges over hundreds of kilometers on the African savannah in search of fresh grazing. It can walk or run for hours at a time and is therefore specialized for endurance rather than for sprinting. However, that is not to say that it cannot run fast when it wants to. Wildebeest can reach top speeds of approximately 80 km/h or 50 mph, compared with approximately only 60 km/h or 37 mph for the lion, which frequently hunts them (see Schaller 1972; Garland 1983). The reason lions are able to capture wildebeest is because they have a much greater acceleration. Provided the lion can get close enough to its prey before making a charge, it stands a good chance of closing the gap and capturing the prey. But the lion, being a sprinter, can keep up these bursts of speed for only short distances.

Among fishes, the pike is a sit-and-wait predator, adapted for rapid acceleration rather than sustained swimming. Its strategy is to wait in ambush for a suitable prey to come within striking distance, then to lunge after the prey, using its large tail for acceleration. The tuna, in contrast, is modified for cruising and spends all its time cruising the ocean in search of suitable prey. Thus the pike is specialized for burst speed, the tuna for endurance. Ducks, as noted earlier, have red flight muscles suited for endurance flying, whereas those of galliform birds are white and are adapted for burst activity. There are corresponding differences in wing structure, with those of game birds being short and broad for rapid acceleration, whereas the slender pointed wings of the duck are suited to endurance flying.

Locomotory Performance and Natural Selection

For a feature to evolve it must be variable and heritable. Do these two criteria hold for speed? Running performance is certainly variable among individuals, as we know from the playing field and racetrack. There is also a long history of the heritability of running performance, as exemplified by racehorses. The correlation between a horse's performance on the track and its pedigree is well established, which is why the owners of successful horses, like the Canadian *Northern Dancer,* are able to command such high stud fees at the end of their horses' running careers. Running performance – the ability to escape predation or to capture prey – is the very essence of survival, and several long-term experiments have been conducted to test it. In one elaborate experiment, the burst speed and maximal exertion (total distance crawled while pursued) of over 600 garter snakes was measured (Bennett 1991). The sample included newly hatched as well as adult individuals. The animals were set free and monitored for the next three years. During their first year of growth, neither burst speed nor maximal exertion were significant survival traits. Instead, body length was a significant factor determining survival, presumably because long snakes were less likely to be preyed upon than short ones. However, during the second and third years, the snakes' burst speed and maximal exertion were both correlated with survival. Other studies have been conducted with reptiles and fishes, sometimes with similar results, but other times with no correlations between speed and survivorship, showing that there were selection pressures that were competing with speed.

Burst speed and endurance speed can both be improved upon by training, as discussed later. To understand why requires some understanding of the physiological differences between the two types of exercise, which can best be demonstrated with the use of a treadmill.

Aerobic and Anaerobic Exercise

If you were sitting quietly at a comfortable room temperature – neither hot enough to make you sweat nor cold enough to make you shiver – and if you had not recently eaten, you could have your basal, or resting, metabolic rate measured. The *basal metabolic rate,* or *resting metabolic rate,* is assessed by measuring the oxygen consumption of the resting animal, and this measurement is often expressed in milliliters of oxygen consumed per kilogram of body mass per hour. (In the more recent literature, metabolic rate is expressed in terms of watts per kilogram of body mass, one watt being the power when one joule of energy is expended in one second.) Metabolic rate is increased by exercise, shivering, sweating, and by digestion, which is why it is important to control the conditions when measuring basal metabolic rate.

Oxygen consumption is usually measured by fitting the subject with a mouthpiece and nose clip, so that the exhaled air can be analyzed and compared with the ambient air to assess the rate at which oxygen is being used by the body. If you were attached to such an apparatus, and the conditions were as mentioned earlier, your basal metabolic rate could be measured. If you now did the slightest amount of exercise, even moving your arms, your metabolic rate would increase slightly. Suppose you now stood on a treadmill and the speed was adjusted to the minimum setting, so that you were walking at a leisurely pace. Your oxygen consumption would start increasing, showing that your metabolic rate was rising above the basal level. After a few minutes your metabolic rate would stabilize at a level higher than the basal rate. If the speed of the treadmill were increased, your oxygen consumption would increase and settle at a new level. If this were repeated, a series of oxygen consumption levels would be obtained. Oxygen consumption therefore increases according to your speed, and this would continue, depending on your level of fitness (in an athletic rather than an evolutionary sense), until you were jogging along at a sustainable level. Under these conditions your skeletal muscles are being supplied with oxygen at the same rate it is being used, and your exercise is entirely aerobic. If a graph of oxygen consumption were plotted against your running speed, a straight line would result. However, if the speed continues to increase, a point is reached when your oxygen consumption no longer increases with increasing running speed. You have now exceeded your *endurance speed,* or *maximum sustainable speed,* the maximum speed at which sufficient oxygen can be supplied to your skeletal muscles to maintain their aerobic respiration. Your lungs and blood vascular system can no longer cope with the increased oxygen demands, and the only way you can continue to go faster is by augmenting the aerobic metabolism of your skeletal muscles with glycolysis. This is the point at which more of the fast fatigable motor units are pressed into service, since they can function anaerobically.

The ratio of an individual's maximum oxygen consumption to his or her resting oxygen consumption is referred to as the *aerobic scope.* Aerobic scope can be increased with training, as anyone who goes jogging knows. This is because you can sustain a higher speed, and therefore a higher oxygen consumption, when you are fit than when you are out of training. The average aerobic scope for humans is about 5; athletes reach levels of about 11, and one remarkably fit Scandinavian skier is reported to have achieved an aerobic scope of 27.5. Once an individual is running above his or her endurance speed, increases in speed are achieved by increasing the level of glycolysis. The maximum burst speed, rather surprisingly, is only about twice as fast as the maximum sustainable, or endurance, speed. I know this is true for myself because I have checked my speeds on a treadmill. You can also check it yourself by consulting the *Guinness Book of Records,* where you will find that top sprinters run only about twice as fast as the top marathon runners.

Amphibians and reptiles have similar aerobic scopes as birds and mammals, but their basal metabolic rate is only about one-tenth as high. Consequently, amphibians and reptiles typically lack stamina and have endurance speeds that are much lower than those of mammals. For example, the basal metabolic rate of a 1 kg iguana would be about 2 ml O_2/minute compared with 9 ml O_2/minute for a 1 kg mammal. Their maximum oxygen consumptions are 9 ml and 54 ml/minute, respectively, giving the iguana an aerobic scope of 4.5 compared with 6 for the mammal, which is about the same. However, whereas the endurance speed of the lizard would be about 0.5 km/h (0.3 mph), that of the mammal would be about 4.1 km/h (2.5 mph; data from Bennett and Ruben 1979). Although reptiles usually have low endurance speeds, most are capable of moving very fast when they want to, as anyone who has ever tried catching a lizard in the sun will know. These high burst speeds, which compare with those of similarly sized birds and mammals, are powered by glycolysis. Since they have such low sustained speeds, reptiles and amphibians have a much greater differential between endurance speeds and burst speeds. Thus, where the maximal running speeds of mammals are generally only about twice that of their sustained speeds, reptiles can increase their speeds by a factor of between ten and thirty. Their lightning increase in speed, as when a seemingly moribund crocodile leaps out of the water to seize an unsuspecting zebra, cannot be kept up for long periods. This is probably attributed to the rapid accumulation of lactic acid, which reptiles seem unable to break down quickly. Just as the buildup of lactic acid in an athlete's muscles causes stiffness and eventual impairment of performance, reptiles become sluggish and even comatose if driven to excess. Some investigators in Australia conducted experiments with lassoed crocodiles to see how long they could maintain their struggles to escape. The smallest individuals succumbed the most readily: those weighing less than 1 kg struggled for only about five minutes before becoming limp and unresponsive, compared with twenty minutes for 10–100 kg individuals and thirty minutes for those over 100 kg (Bennett et al. 1985). The lactic acid levels measured in the blood of the exhausted crocodiles were the highest ever recorded for any animal, and it took the reptiles about two hours to make a partial recovery.

Being an empiricist I conducted a similar experiment, though the only reptile readily available to me at the time was a small lizard that was kept in our fossil preparation laboratory. I harassed the lizard with my finger for several minutes, after which time it became completely unresponsive. I assured everyone in the lab that it would recover within the hour, but to my shame, it never did. I am still occasionally reminded of the dastardly deed.

Because of their general lack of stamina, most predatory reptiles use a sit-and-wait strategy for hunting rather than actively searching out their prey like most mammals and birds. Many snakes, for example, spend hours, even days, coiled up inconspicuously in an appropriate location, waiting for a suitable prey to come along. Some snakes do actively search for prey, though, slithering down burrows, climbing trees, and looking in other likely places. They usually do not travel very far during these forays, nor do they move at anything but a leisurely pace, but there are some interesting exceptions. The racer (*Coluber constrictor*) and coachwhip (*Masticophis flagellum*), for example, both native to North America, are fast, slenderly built snakes that actively hunt for prey, relying on their speed to avoid being preyed upon themselves. At the other end of the activity spectrum are heavily built and relatively slow-moving snakes like the rosy boa (*Lichanura roseofusca*), also native to North America, a sit-and-wait predator. A comparison between the activity levels of the rosy boa, coachwhip, racer, and the western rattlesnake (*Crotalus viridis*) was made by stimulating the snakes into vigorous exercise for a five-minute period (Ruben 1976, 1979). The rosy boas did not remain active for the entire five minutes, spending some of the time coiled up in a defensive ball. The rattlesnakes kept active but were completely exhausted at the end of the period. The racers and coachwhips showed no signs of exhaustion and were not completely exhausted even after another five minutes of exercise. More than half of the energy used by the snakes during their forced activity was provided by anaerobic metabolism. But the racers and coachwhips had the advantage over the other species in generating much higher levels of aerobic energy – far higher than the other snakes could generate anaerobically. Part of the reason for the racers and coachwhips having so much stamina is that they have more complex lungs. These lungs are better supplied with blood vessels, allowing them to extract more oxygen from the air.

Other snakes also appear to have lots of stamina, like the sidewinder rattlesnake (*Crotalus cerastes*). Sidewinders can travel well over 1 km during a single night's hunting. This is an impressive performance by reptilian standards, but it is insignificant compared with the 40 km hunting forays of African wild dogs (*Lycaon pictus*). Far more impressive are cobras, which become territorial during the breeding season, guarding their domain with considerable zeal. They raise their heads high off the ground to peer over the top of the undergrowth, and human intruders have reportedly been chased over distances of many yards and been bitten.

Perhaps the most active of all reptilian predators is the Komodo dragon (*Varanus komodoensis*). This large monitor lizard reaches lengths in excess of 3 m (10 feet), but most individuals are much smaller. They are active scavengers, roaming over large distances in search of food. Adult males travel as far as 10 km (6 miles) in a day, though the average is closer to 2 km. They usually move fairly slowly, about 4.8 km/h (3 mph), which is about half the speed of a foraging wild dog. But they can move much faster, and startled individuals have been clocked running at speeds of 14–18.5 km/h (9–11 mph). One large individual (2 m; 6.5 feet) was followed on a motorcycle at a speed of 14 km/h (9 mph), which it kept up for about two minutes. It is also reported that Komodo dragons can run at speeds approaching 30 km/h (20 mph) for distances of about half a mile (1 km; Auffenberg 1987).

These examples show that there is a fairly broad spectrum of activity levels among reptiles and that not all of them are sluggish, sit-and-wait predators. However, most reptiles are fairly inactive compared with most birds and mammals, and they are far more reliant on anaerobic metabolism for powering their rapid activities. Before discussing the possible evolution of stamina, I would like to digress and say something about the differences between training for endurance events, like long-distance races, and for sprints, like the 100-meter event.

Training for Endurance and for Speed

My old school, being very English, had the tradition of an annual cross-country run. I was never very good, and always finished well after the winner had crossed the line. We used to try to guess the likely winner, but some of the obvious choices, like the big muscular lad who worked out with weights, seldom did well at all. The winner always seemed to be the skinny kid who nobody would have picked as a finisher, let alone a victor. You may have noticed the same thing when watching the Olympics: marathon runners and long-distance runners are generally of slight build, whereas sprinters have the physiques of bodybuilders.

The reason for the discrepancy is that long-distance runners are trained for endurance, their event being performed aerobically, whereas 100-meter sprinters run most of their race anaerobically. It does not matter to a sprinter whether his or her muscles are adequately supplied with oxygen because most of the sprinter's muscle power is anaerobic. What matters to a sprinter is acceleration, which is proportional to the force generated by the muscles (force = mass × acceleration). As seen earlier, the force generated by a muscle is proportional to its cross-sectional area; therefore, sprinters need large leg muscles. Large leg muscles require a large skeleton and also a strong upper body to reduce twisting during acceleration, hence the sprinter's broad chest. Sprinters use high-intensity resistance training to prepare for their event, making much use of weights. This workout increases the size of their muscles (skeletal muscles increase in size by increasing the diameter of individual muscle cells). Long-distance runners, in contrast, use endurance exercises for their training, such as swimming, cycling, and long-distance running. This kind of exercise has a number of effects on their blood vascular system. These effects include increasing the size of the heart, thereby increasing the volume of blood pumped during each cycle, and increasing the capillary supply to the muscles. The density of mitochondria in the slow-oxidative muscle cells (SO) increases, and some fast cells may be converted into SO cells. The individual muscle cells do not undergo significant enlargement, though some of the leg muscles (for example, the anterior tibial muscle in the shin) may increase in size a little.

The Evolution of Stamina

Fishes have low metabolic rates, similar to those of amphibians and reptiles. However, since transport costs in water are only about one-tenth of those on land, fishes are capable of endurance performance. Tunas, for example, like many large pelagic fishes including marlins and sharks, swim continuously, maintaining speeds in the order of about 4 km/h (2–3 mph; see Block and Booth 1992; Carrier 1987). The first terrestrial vertebrates, early amphibians that lived over 350 million years ago, inherited low metabolic rates from their fish ancestors. Faced with the increased transport costs of moving over land rather than through water, they probably had much lower endurance speeds than their swimming ancestors. These low endurance speeds appear to have been retained by modern amphibians, and the sustainable walking speeds of modern salamanders are less than one-tenth those of the swimming speeds of similar sized fishes (Bennett 1991).

With a few notable exceptions (like frogs and chelonians), amphibians

Figure 9.5. The sprawling posture of amphibians and reptiles (left) contrasts with the erect posture of birds and mammals (right).

and reptiles undulate their bodies when they walk, a feature they inherited from their piscine ancestors. Flexing the body from side to side tends to interfere with pulmonary respiration, because the lung on the flexed side of the body tends to be collapsed. D. R. Carrier (1987) proposed that this flexing was a constraint to the evolution of locomotory stamina in amphibians and reptiles. Birds and mammals overcame this constraint by abandoning the sprawling posture and lateral body movements of their tetrapod ancestors. Instead they evolved an erect posture where the legs moved fore and aft rather than from side to side (Fig. 9.5). Although this is a most plausible scenario, it is not completely satisfactory because, as we have seen, there are several reptiles, including snakes like the coachwhip, whose lungs obviously do function perfectly adequately during their undulatory movements. It seems likely that a number of factors have contributed to the evolution of stamina in birds and mammals, including their erect posture and the physiological changes correlated with the evolution of thermoregulation.

PRACTICAL

The purpose of this lab is twofold: to introduce you to the complexity of the skeletal muscular system of vertebrates; and to familiarize you with the mechanics of simple levers. There is a lot of material to get through, and the work should be spread over two practical sessions.

PART 1 THE DISSECTION OF THE LEG MUSCLES OF THE PIGEON

Space and time do not permit this lab outline to be a dissection guide to a bird, but a few general pointers will be given. We will first be concerned with the muscles of the lower leg (from knee to ankle); the illustrations should permit you to find and identify the relevant muscles (the dissections illustrated pertain to the kiwi, but there are only minor differences from other birds and these should not concern us). After dissecting these muscles, we will dissect the knee joint.

CAUTION: The material you are dissecting is fresh, therefore there is a risk of infection should you cut yourself (or if you have any areas of broken skin on your hands). If you should cut or prick yourself while handling this material:

1. Make the wound bleed to wash out the bugs.
2. After bleeding apply alcohol liberally.

To reduce these risks, wear rubber gloves.

When you have finished dissecting for the session, wash your hands and instruments liberally with soap and water. Wipe your instruments liberally with alcohol and dry them off. As a precaution I also wipe my hands off with alcohol, finishing off with soap and water.

Dissection, careful dissection, takes time. It will take several hours to complete these dissections, which usually continue over two labs. Note that muscles have Latin names and are prefixed with the upper caseletter M, standing for the Latin word for muscle.

Procedure

Selecting the left leg, carefully remove the feathers so that the skin is not torn. There is no need to go much further up the leg than the level of the knee. Remove the skin from the plucked area, taking care not to remove the underlying muscles.

Carefully examine the muscles and see whether you can identify those shown in Figure 9.6. The individual bellies of the muscle are enclosed in thin transparent sheets of connective tissue, which are all attached to one another. Using blunt dissection (that is, forceps and seeker), separate the individual muscles. You must take great care in separating one muscle from another – it is all too easy to "invent" a new muscle by splitting part of a single belly away from the rest of the muscle. It is usually best to identify the tendon of insertion of a given muscle and work toward the origin. The connective tissue joining adjacent muscles is usually thin enough to be broken with the tip of the seeker (probe), but if it is necessary to cut through it, do so very carefully, using fine-tipped scissors and making sure that you can see precisely where you are cutting (it is best to use the seeker as a guide). Muscle bellies should part cleanly, one from another. If you see lots of ragged ends where muscle fibers have been separated, you should ask yourself whether you have not broken a single muscle in two. When you have cleared (separated) the muscles – you should be able to run your seeker behind them – determine, as well as you can, the point of origin and insertion of each muscle.

Assess the action of each muscle by pulling gently on its tendon of insertion. Do this with the ankle joint partly flexed. Remember, flexion is when the angle between the two segments of a joint is decreased, that is, the joint is closed. Extension is when the angle is increased, straightening out the lower segment. Try to

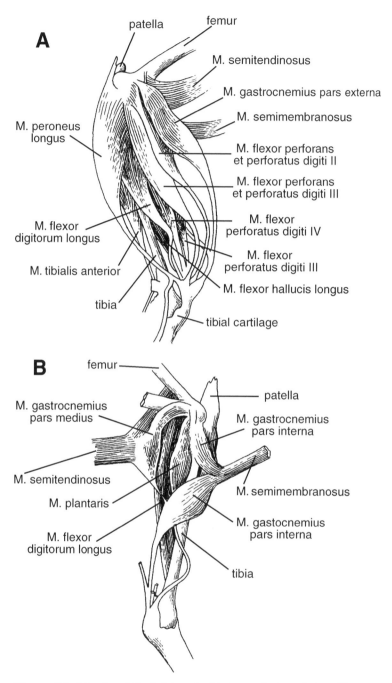

Figure 9.6. Muscles of the left leg of a bird, seen in lateral view (A) and in medial and somewhat caudal view (B). This illustration gives some idea of the complexity of the arrangement of muscles about a joint.

detect which way the lower segment wants to move. Is the muscle a flexor or extensor? List the muscles you identify, noting their origin, insertion, and whether they are flexors or extensors.

You will find that several of the muscles have a similar action. Do similarly acting muscles necessarily have the same mechanical advantage? The M. peroneus longus is an extensor of the intertarsal joint, so too are the Mm. flexor perforans et perforatus digiti II and III (they are called flexors because they flex the toes – specifically toes II and III – but they also bring about an extension of the intertarsal joint) and the M. gastrocnemius pars externa. If you give the tendons of insertion of each of these muscles a small pull, you will see that the action is far less powerful in the M. peroneus longus than it is in the other muscles. This is because the mechanical advantage of the M. peroneus longus is much less than that of the other muscles. Why do the other muscles have greater mechanical advantages? If you examine your specimen, comparing it with Figure 9.6, you will see that the tendon of insertion of the M. peroneus longus lies close to the pivot, whereas the tendons of insertion of the other muscles attach to the tibial cartilage. Since the latter is a little farther from the pivot, these other muscles have longer moment arms and hence greater mechanical advantages.

Do muscles with similar actions necessarily function in the same way? If you dissected out more muscles you would find that some muscles were more effective in operating a particular joint when it was partly flexed, and others were more effective when the joint was more extended. Similarly, some muscles bring about a greater excursion of a joint than others. Thus, although it appears that there is some redundancy in muscles – there are more muscles present than appear to be needed for the job – they do not necessarily all have exactly the same function. Having several muscles that do slightly different, though essentially similar, jobs probably allows an animal to make more precise movements. There is no question that animals can locomote quite satisfactorily with some of their muscles out of action, though they may be lacking in their repertoire of fine adjustments. For example, pigeons that have lost the use of the M. supracoracoideus, one of the main flight muscles, which elevates the wing, can still fly satisfactorily. However, they have difficulty in taking off.

Fiber Orientation Locate the M. gastrocnemius pars externa, the large posterior calf muscle. Locate the tendon of insertion and follow its course over the posterior surface of the belly, where it widens into a broad aponeurosis (sheet of connective tissue). Locate the aponeurosis of origin on the opposite surface of the belly. The muscle fibers run unipennately between these two aponeuroses. To see this, you will have to cut a longitudinal section through the belly of the muscle. This is best done using a pair of scissors while the muscle is still in situ.

PART 2 INVESTIGATING THE ACTION OF LEVERS

We are now going to perform some simple experiments that are important in interpreting certain aspects of living skeletons.

You are provided with a tongue depressor drilled with holes at 10 mm

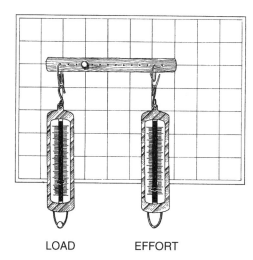

LOAD   EFFORT

Figure 9.7. A simple apparatus for assessing the action of levers.

intervals, a pair of spring balances, a wooden board, a G-clamp, some pins, some paper clips, and graph paper. Set the lever up as shown in Figure 9.7, making sure that the pin is firmly anchored in the board. Clamp the board to the top of the table using the G-clamp. The spring balances should read zero; if not, you can adjust them by pulling or pushing on the metal tag that is part of the scale. These balances are neither very accurate nor very precise, and they are the source of much of the experimental error you will encounter.

The right-hand balance, arbitrarily designated the effort, is attached nine spaces (90 mm) from the fulcrum. The left-hand balance, the load, is attached one space (30 mm) from the fulcrum. Recall that the effort is the force applied to the lever, whereas the load is the resultant force. If the fulcrum (joint in the skeletal system) were frictionless, and if the energy converters (muscles in the skeletal system) were 100 percent efficient, the work put into the system (effort × distance moved by effort) would be equal to the work output of the system (load × distance moved by load). In our simple system we will see that the work put into the system is roughly the same as the output. In the living situation, the efficiency of muscles is far less than 100 percent; therefore, the amount of energy that has to be supplied to a muscle is much more than the amount of work that muscle performs during a contraction.

Experiment 1   Investigating the relationship between effort and load with a simple lever system.

Making sure that the lever remains horizontal, take up the slack on both balances – they should register no load, or just a small one. Mark the position of the effort and load by pricking the graph paper through the holes where the paper clips attach. Anchor the *load* balance with one of the pins, making sure that it is perpendicular to the lever. Pull down on the effort balance until a reading of 100 g

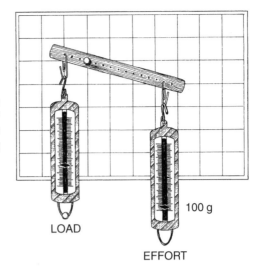

Figure 9.8. The force on the effort is increased in 100 g steps, and the new positions of the effort and load are marked with pinpricks on the underlying graph paper.

100 g

LOAD

EFFORT

is reached, making sure that its line of action is kept perpendicular to the original (horizontal) position of the lever. You can confirm that it is perpendicular by lining up the effort balance along the vertical lines on the graph paper. The lever will now be at an angle (Fig. 9.8). Mark the new positions of the effort and load by puncturing the graph paper through the two paper clip holes, as before. Record and tabulate the readings on the effort and load balances.

Repeat this procedure, increasing the effort by 100 g increments. You will be able to keep the line of action of the effort at right angles to the original position of the lever, but not that of the load because it is fixed. The discrepancy is not very much for the smaller efforts, but you will notice that by the time the effort is more than 400 g, the line of action of the load is well off the vertical. This discrepancy will affect the results, but it is an unavoidable shortcoming of the experimental design. The results will still give you a good idea of what is going on.

Do not attempt to exert more than about 500 g on the effort balance. Remove the lever and locate the pinpricks. Join up the corresponding effort and load position points as shown in Figure 9.9. Measure the moment arms of the effort and load for each position. Also measure the distance through which the effort and load have moved after each increment. You can measure this distance along the vertical axis of the graph paper, from the original position of the pinprick to the new position. Tabulate the following results: effort (g), load (g), moment arm of effort (mm), distance effort moves (mm), load, moment arm of load, and distance load moves. From these results, calculate the following:

1. Mechanical advantage = force applied to load (F_l)/force applied by effort (F_e)

Strictly speaking, F_l and F_e should be expressed in newtons, as the product of mass (kg) and the acceleration due to gravity (9.81 m/s/s). However, since F_l/F_e is a dimensionless ratio, this is not necessary; you can simply express the ratio in

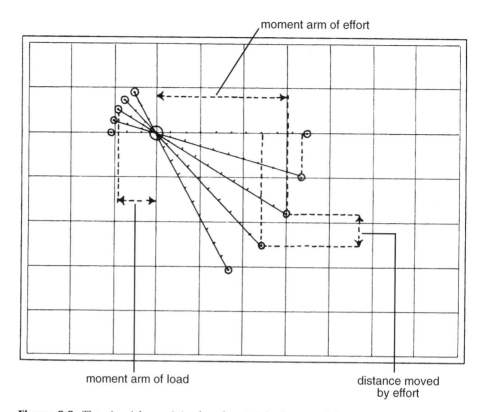

Figure 9.9. The pinpricks are joined up by straight lines, and the moment arms and distances moved by the efforts and loads are measured.

terms of the readings, in grams, on the load and effort balances. The mechanical advantage gives a measure of how much force a given effort can produce. This means that an input force (effort) of 10 newtons would give an output (load) of 30 newtons. However, the cost of getting all this extra force is that the effort has to move three times as far as the load (compare the distance the effort moves with that moved by load). Thus the *work* obtained from the system is the same as the work put in, or nearly so. You will find that the mechanical advantage of this system is about 3. Theoretically it should be exactly 3, but it will be significantly less for several reasons. First, the line of action of the load departs from being vertical to the original position of the lever. This discrepancy increases with increasing angles of the lever. Second, significant measurement errors can be ascribed to the spring balances. There is also some energy lost through friction, and the work got out is therefore always less than that put in.

2. Moment arm of effort/moment arm of load.

This number will have remained constant throughout the experiment, at about 3. If you compare this ratio with the previous one, you will find they are numerically equal (within experimental error). Thus, mechanical advantage = moment arm of effort/moment arm of load.

3. Velocity ratio = distance effort moves/distance load moves.

The velocity ratio of this system is about 3. This means that in order to move the load through a distance of 1 cm, the effort has to move through 3 cm. Note that the velocity ratio is numerically equal to the mechanical advantage.

4. Work done by effort and work done on load.

To obtain a value for the work done, in joules, you would first need to calculate each force, in newtons, by multiplying the balance reading, in kilograms, by 9.81. These values would then have to be multiplied by the distance moved by the forces, in meters, to obtain work, in joules. Since all balance readings would be treated the same, you can eliminate the tedious and unnecessary duplication simply by calculating "work" as the product of the balance reading, in grams, by the distance moved, in millimeters.

The lever system you have been assessing is a class 1 lever, so designated because the effort and load act on opposite sides of the fulcrum. In this particular case the lever has a mechanical advantage in excess of one, and such levers are used in applications where large forces are required. Because of the high velocity ratio (effort moves through a larger distance than the load), the action is relatively slow. Such slow, powerful movements are common in physical systems (e.g., nutcrackers, a screwdriver being used to pry open a paint can) but not in most animals. The exceptions to this generalization are fossorial animals (burrowers and diggers) like moles and anteaters, whose limbs move more slowly but with greater forces. The usual situation in animals is to have low-velocity-ratio levers. If you reversed roles and designated the left-hand balance as the effort and the right-hand balance as the load, the lever system would have a low mechanical advantage and a low velocity ratio, which is more like skeletal systems. As noted earlier, the forearm is one such system: the triceps muscle inserts close to the pivot (elbow joint), and a small contraction in the muscle causes a large displacement of the hand.

Experiment 2   Assessing the lever system of the human forelimb.

Examine the articulated skeleton of the human forelimb. Notice the points of origin and insertion of the biceps muscle (Fig. 9.10 marked by crosses). Hold the limb so that the radius and ulna are perpendicular to the humerus. Imagine a weight in the palm of the hand and the effort that the biceps needs to exert in order to lift the weight. Measure the moment arm of the biceps. Deduce the mechanical advantage and velocity ratio of the biceps muscle when raising a weight.

Holding your own arm in this position, determine the maximum weight you are able to lift. Calculate the force exerted at the palm (in newtons) when the weight is just lifted clear of the tabletop. What force has to be developed in the biceps muscle to lift the weight?

Experiment 3   Investigating the effect of inclining the line of action of the effort.

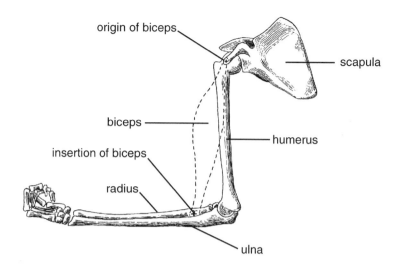

Figure 9.10. The skeleton of the human forelimb and its relationship with the biceps muscle.

Repeat the setup as in the first experiment, but attach the fulcrum pin close to the top left-hand corner of the graph paper. Keeping the lever horizontal, and the two balances perpendicular to it, apply forces to the two balances until the left-hand one (load) registers 1 kg. Because the mechanical advantage of the system is 3, the effort required to balance this load is only 1/3 kg. Anchor the load (left-hand balance). Taking care to keep the lever horizontal, incline the effort progressively more toward the horizontal (swing it toward the right) until the effort is equal to the load (Fig. 9.11). Measure the moment arm of the effort. It takes a little imagination to do this because you have to draw a line corresponding to the line of action of the effort, extrapolating beyond the lever. The moment arm of the load has not moved; it is still about 30 mm.

You should have found that the (new) moment arm of the effort is the same as that of the load. The mechanical advantage of this new system is one ($F_l = F_e$, moment arm of effort = moment arm of load). This experiment emphasizes the point that moment arms are always measured at right angles to the lines of action of the forces.

In the two lever experiments you conducted, the effort was applied at the point farthest from the fulcrum, which is not typical of skeletal situations, as already noted. The reason why the experiment was set up like this was to minimize the forces on the rather flimsy apparatus. For the last experiment we will designate the balance closest to the pivot as the effort, as in the typical skeletal situation. Be aware that the forces exerted may destroy the apparatus.

Experiment 4   Investigating a low-mechanical-advantage lever system where the line of action of the effort is inclined.

Set up the lever system as in the first experiment, but designate the left-hand balance as the effort. Taking care to keep the lever horizontal, take up the slack in

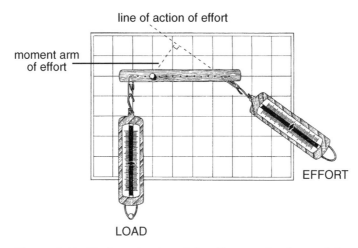

Figure 9.11. The line of action of the effort is inclined toward the horizontal.

both balances and anchor the load balance (right). Swing the left-hand balance (effort) to the left so that its line of action is inclined at an angle of about 45° to the lever. Apply a force to the left-hand balance until it registers 1 kg. What is the reading on the right-hand balance? You will find that the effort is more than four times higher than the load.

Many of the muscles that close the jaws (adductors) have a line of action that is inclined to the horizontal axis. Large forces have to be exerted by these muscles to produce relatively small load forces. One of the main reasons for this seemingly wasteful arrangement is to keep the jaw firmly pressed into its articular socket, thus preventing jaw dislocation when chewing. Imagine that the lever is a jaw and the fulcrum the hinge joint (the "skull" is closest to you). While maintaining the effort, loosen the fulcrum. Notice how the "jaw" is pulled in toward the "skull."

Terrestrial Locomotion

We are somewhat unusual among tetrapods (four-footed animals – amphibians, reptiles,[1] birds, and mammals) in being bipedal, a feature we share with kangaroos, birds, and many dinosaurs. Most other terrestrial vertebrates locomote on all fours. Almost all reptiles and amphibians, as mentioned in the previous chapter, have a sprawling gait and an undulatory way of moving their bodies when they walk. Most mammals, in contrast, have an erect posture, where the fore and hind legs are held beneath the body and the leg bones are held vertically when viewed from the front or from the back. This chapter is largely concerned with locomotion in quadrupedal mammals, though some of the observations and experiments will be conducted upon our bipedal selves. Before starting, though, I need to define two terms. *Stride length* is the distance between footfalls of the same foot. For example, if you walked through a puddle, leaving your footprints behind, the distance between the impression of your left foot and the next print of your left foot is your stride length. If you started running you would find that your stride length increased considerably. When I walk at a leisurely stroll, my stride length is slightly more than a meter, but it increases to about three meters when I am sprinting. *Duty factor* is the fraction of time in a stride cycle that a particular foot is on the ground. Suppose it took you 3 seconds to complete one stride when you were walking slowly. If your right foot were on the ground for a total of one second of that time, its duty factor would be 1/3, or 0.33. The duty factor of a given foot decreases as speed increases.

Walking, Trotting, and Galloping

Suppose a horse were walking down the road accompanied by a large dog and small puppy, all moving in line abreast at the same speed. The horse is walking, the dog is trotting, but the puppy is having to gallop to keep up with its larger consorts. The walk, the trot, and the gallop are three distinct patterns of locomotion, called *gaits,* that quadrupeds use to move at increasingly higher ranges of speed. The three gaits are analogous to the gears of a car. At slow speeds quadrupeds walk, increasing their speed by increasing the number of strides they take per minute. A graph of speed (*x* axis) plotted against

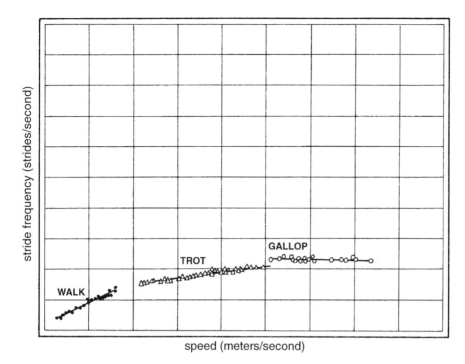

Figure 10.1. Stride frequency increases with speed during the walk and trot but remains essentially constant during the gallop. (Data pertain to the horse.)

stride frequency (y axis) would give a straight line with a positive slope. As stride frequency (revs) increases, a point is reached when an animal can no longer go any faster without changing gait to a trot. The animal "changes gear," but the car analogy is not perfect because there is no drop in revs. Instead, stride frequency continues to increase, though more slowly, so that the speed/frequency graph has a lesser slope (Fig. 10.1). The reason why changing gait from a walk to a trot increases the speed range is because it increases stride lengths. Thus the stride length of a trot is greater than that of a walk, at a particular stride frequency. Stride frequency continues to increase with increasing speed throughout the trot, until the animal changes to the gallop. Unlike the walk and the trot, the gallop produces only a slight increase in stride frequency; the higher speeds are instead achieved by increasing the length of the stride. A graph of stride frequency plotted against speed would therefore have a shallow gradient (Fig. 10.1). When an animal changes from a trot to a gallop it has reached its limit for making rapid body movements and is essentially running at its maximum sustained stride frequency. It has been proposed that the transition from the trot to the gallop – the point at which stride frequency cannot be significantly increased – represents a physiologically similar speed among quadrupeds (Heglund, Taylor, and McMahon 1974). A zebra and a fox that are running at the trot/gallop transition are therefore said to be running at the same equivalent speed, even though the zebra is traveling considerably faster in terms of miles per hour. This concept permits us to make comparisons between the locomotory performances of animals of different sizes.

The speed at which an animal changes gait increases with increasing body size and has been documented for a wide range of animals (Heglund and Taylor 1988). This phenomenon explains why small children have to jog to keep up with the adult walking by their side. The speeds attained at each gait also increase with increasing body size. Therefore if the horse, the dog, and the puppy in our earlier example were all walking along at a comfortable pace, the horse would be way ahead of the dog, and the puppy would be left far behind. The same would also hold for the trot and the gallop. However, stride frequency decreases with increasing size. A walking puppy therefore takes more strides per minute than a walking dog, which in turn takes more strides per minute than a walking horse. Smaller animals, then, are relatively slower and have higher stride frequencies than larger ones.

Another way of looking at the relationship between body size and gait changes involves the use of the Froude number. This dimensionless number, originally derived for studying ships, has a wide range of applications and therefore deserves attention.

Froude Number

In most nautical movies, especially the older ones, it is usually easy to spot the sequences that were shot in a tank because the models move with an unnatural jerkiness relative to the water. This motion is of minor consequence in terms of entertainment, but the relationship between the movements of a model and of a full-sized ship is of considerable importance to naval architects because designs are initially tested on models. William Froude (1810–1879), a British engineer and ship designer, sought a way of scaling models so that their behavior in the water would be equivalent to that of a full-sized ship at sea. Two forces, inertial and gravitational, dominate a ship's performance, and Froude combined these forces into a ratio that now bears his name. The Froude number (Fr) is given by $Fr = v^2/gl$, where v = velocity, g = the acceleration due to gravity, and l = length.

One implication of this relationship is that if a model is built with the same Froude number as the real ship, the wave patterns caused by the movements of its hull relative to the water will be the same as in real life. For example, suppose the model is 1/100 as long as the real ship, and that the latter moves at 20 mph. Provided the model moves at 2 mph, its movements will be just like those of the full-sized vessel. The Froude number has numerous applications. For example, it has been found that the maximum speed that is practical for a ship to travel approximates to a Froude number of about 0.16. That means that large ships can go faster than smaller ones.

Although Froude numbers are primarily used for things that move in water, a novel departure has been made in an application to terrestrial loco-motion (Alexander 1976, 1989; Alexander and Jayes 1983). The rationale for this application was that the movement of the leg, which was equated to that of a pendulum, and the footfalls of the foot are largely under the influence of gravity (Alexander 1984b). Length in the equation for the Froude number was taken as the distance between the hip joint and the ground, and velocity was taken as the speed of the animal. One of the implications of this work is that each gait has an optimum Froude number range so that the transition from one gait to another should occur at approximately the same Froude number; this theory has been verified experimentally. Another interesting prediction was that running speeds could be estimated from trackways, given the ani-mal's hip height. This idea was based on some experimental evidence for a diverse array of mammals, where a graph of Froude number plotted against stride length gave a straight line (Alexander 1976). This concept then led to the use of Froude numbers for estimating the running speed of dinosaurs from their fossilized trackways, based on estimates of their hip heights. For reasons discussed elsewhere (McGowan 1991), which include the fact that the scatter in the data points became much wider as more species were sampled, the methodology has serious limitations. Nevertheless, it continues to be used by many paleontologists, resulting in a large data set of dubious estimates for dinosaur speeds.

Before we return to the subject of gaits with a description of the differ-ences between the three types, it is necessary to consider the role of the cen-ter of mass in posture and locomotion. The *center of mass,* or *center of gravity,* of a body is the point where the entire mass can be considered as being located. If a body is supported at its center of mass, all the forces acting upon it are in equilibrium, and this is therefore its balance point. I can balance a book on my fingertip if I place it beneath the book's center of mass. Similarly, a quadruped can balance its body on three legs, provided it keeps its center of mass over the top of the triangle formed on the ground between its three feet.

The differences between the three gaits are largely a matter of differences in stability and in the degree of oscillation of the animal's center of mass. When a quadruped is walking slowly, it keeps three of its feet on the ground all of the time (Fig. 10.2). Provided its center of mass remains inside the trian-gle formed on the ground between the supporting legs, it remains stable and does not tend to topple when swinging its fourth leg forward. As stride fre-quency increases, stability is sacrificed for speed, and no more than two feet are on the ground at any instant. The animal tends to topple toward the unsupported legs, causing more displacement of its center of mass. The trend is carried further in the trot because the duty factor of the two supporting legs decreases, and there are very brief moments when none of the feet are on the ground. The gallop is characterized by an extension of the unsupported phase, during which time the animal launches itself into the air, displacing its center of mass even further.

Figure 10.2. The walk (left), trot (middle), and gallop (right).

Horses have stiff backs, and their trunk muscles play a lesser role during locomotion than they do in many other animals, such as dogs and lions, and probably act largely in stabilizing the skeleton against twisting. Animals with flexible backs derive considerable locomotory effort from their trunk muscles. In these animals the vertebral column flexes and extends like a spring, driving the animal forward and extending its stride length (Fig. 10.3).

Storing Strain Energy

In addition to actively contributing to locomotion by supplying muscular energy, the vertebral column also makes a passive contribution through the storage of strain energy (Alexander, Dimery, and Ker 1985). The major component of this mechanism involves an extensive aponeurosis of one of the dorsal back muscles. Tendon, as we have seen, has a low Young's modulus (Chapter 3) and therefore has the potential to store large amounts of strain energy (Chapter 4). It has been suggested that this back tendon accounts for about 70 percent of the total strain energy stored in the vertebral column, whereas the muscle contributes about 10 percent and the elasticity of the vertebrae themselves accounts for about 20 percent.

When you examine a dry-mounted skeleton it does not give the impression of flexibility – not like the supple arch of a bow carved from willow. I always had difficulty visualizing how a skeleton could store very much strain energy, but that was before I dissected Tantor the elephant. Once we had

Figure 10.3. Animals like cats derive considerable locomotory effort from the springing action of their vertebral column.

stripped the carcass down to the bone I was amazed at the flexibility of the skeleton. Pushing down on the ends of the ribs was reminiscent of flexing a saw blade, and the whole skeleton gave the impression of flexibility.

Many of the subtleties of animal locomotion are lost to us because they happen so quickly, but we can overcome the problem by looking at slow-motion films of animals in motion. A film of a cursorial animal like a horse trotting or galloping would show the feet snapping backward as they leave the ground. The snap is largely brought about by the springing action of tendons at the back of the leg. These tendons get stretched when the hoof strikes the ground, and they continue being stretched as the leg flexes under the animal's weight. The stretched tendons consequently store large amounts of strain energy. This strain energy is released during the push-off phase, when the horse is pushed upward and forward. The residual strain energy causes the hoof to snap back as it leaves the ground. This recoil action allows much of the kinetic energy of the footfall to be recovered, significantly reducing the cost of locomotion. This mechanism was described for the horse half a century ago (Camp and Smith 1942) and has been fairly intensively investigated since, not only for the horse but for many other animals, too (Alexander 1984c; Dimery, Alexander and Ker 1986).

Humans are not very good runners, and although we lack the recoil mechanism seen in cursorial animals, we do have some ability to store strain energy. We store energy in the Achilles tendon of the calf and in the ligaments of the arch of the foot, both of which are flexed during each step. It has been estimated that these two structures can store about 35 percent and 17 percent, respectively, of the total energy turnover during each stride (Ker et al. 1987). This storage capacity explains why the shoes worn by competitive sprinters are so thin. Jogging shoes, in contrast, are well cushioned to absorb impact and reduce injuries, but sprinters do not wear them because they dampen much of the springing action of the arch and Achilles tendon. Before leaving the topic of locomotion, we need to consider the biological significance of the relative placement of limb muscles.

Muscle Placement and Inertia

For many years now it has been recognized that in many cursorial animals – horses, antelopes, ostriches, and the like – the bulk of their leg muscles are concentrated proximally, that is, closest to the body, their action being transmitted to the lower parts of the legs by long tendons. This arrangement has traditionally been interpreted as a strategy for reducing the energy costs of locomotion by reducing the inertia of the legs. The inertia of an object is its tendency to stay where it is. That the inertia of a moving object, like a limb, can be reduced by concentrating its mass proximally can be demonstrated by a simple experiment. Take a drinking straw and wrap a piece of modeling clay around it, close to one end, and smooth it down so that it is firmly attached.

Skewer the end of the straw closest to the blob with a pin so that you can swing the straw back and forth like a pendulum. You will find that it takes very little effort to swing it. However, if you repeat the experiment with the pendulum pivoted from the other end, more effort is required to keep it moving. This is because the inertia of the system has been increased by placing the mass distally.

Animals with proximally placed leg muscles are therefore expected to expend less energy in running. However, when some experiments were performed on such animals, they were found to do little better than animals with more distally placed muscles (Taylor et al. 1974). Indeed, the costs of running at the same equivalent speeds are remarkably constant over a wide range of animals, both quadrupeds and bipeds (Heglund and Taylor 1988). How could these rather unexpected findings be explained?

The energetic costs of locomotion have two major components: there are muscular costs – the costs of activating the muscles and of getting work out of them; and mechanical costs – the costs of moving the various parts of the limbs relative to the rest of the body, moving the legs relative to the ground, and raising and lowering the center of mass. The muscular costs are determined primarily by the amount of muscle present, which is obviously related to body mass, whereas the mechanical costs mainly relate to the anatomy of the body. It has been argued that the muscular costs are the predominant ones (Heglund and Taylor 1988), so the anatomy of an animal may have less relevance to the energetic costs of locomotion than previously thought. Be that as it may, the conclusion that the relative placement of limb muscles appears to have nothing to do with costs of locomotion is difficult to accept.

One of the problems with the experiments conducted by C. R. Taylor and his colleagues was that the animals being compared (cheetah, goat, and gazelle) differed from one another in more regards than just their limb muscles. What was needed was an experiment where the only variable was the relative placement of the muscle mass. Such an experiment, of course, is not possible, but experiments have been performed on humans where the energetic costs of running were compared for individuals before and after having a load attached to various parts of their bodies (Myers and Steudel 1985). In this carefully designed series of experiments, an individual's energetic costs while running without a load were compared with those when a fixed load was attached to various parts of the body: the waist, thigh, shin, and ankle. The costs increased above the unloaded condition by 4, 9, 12, and 24 percent, respectively. These results clearly demonstrated that the energetic costs of running increased as weight was added to progressively more distal portions of the limb. Similar conclusions have been reached by other researchers using an entirely different approach (Hildebrand and Hurley 1985).

Trained sprinters tuck in their legs during the recovery stroke such that their heels almost touch their buttocks; this movement can be seen during slow-motion reruns of races. The marked reduction in inertia of the leg speeds

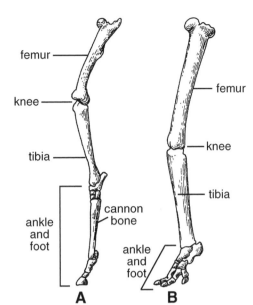

Figure 10.4. The hind limbs of a horse (A) and elephant (B). Notice how much longer is the lower leg segment (below the knee) in the horse than in the elephant.

the recovery stroke, thereby increasing the stride rate and hence the speed. Runners are not trained to do this consciously, but their training includes stretching exercises that increase the mobility of the femur, especially the extent to which it can be retracted (drawn back). The tuck-in therefore comes naturally to them.[2] Joggers can simulate this motion for themselves simply by making a conscious effort to tuck in the lower leg as soon as the foot leaves the ground. If this is done in the middle of normal running, an immediate acceleration in stride rate is experienced. This results in a burst of speed, but the speed of the power stroke also has to be increased in order to sustain the increase in stride frequency.

The bulk of the muscles that swing the leg back and forth during locomotion insert upon the upper limb bones, the femur and humerus (the propodials). It therefore follows that increasing the length of the lower leg segment – from the knee and elbow down to the foot – relative to the upper segment decreases the mechanical advantage, and the velocity ratio, of the system. This means that a given contraction of the muscles will cause a greater displacement of the distal end of the limb, thereby increasing the speed of the foot. Conversely, having a shorter lower limb segment relative to the upper segment gives a slower, more forceful movement of the leg. A comparison between the legs of a fast runner, like the horse, and a noncursorial animal, such as the elephant, which is graviportal, shows this point very convincingly (Fig. 10.4A, B). Part of the extension of the horse's lower leg has been achieved by elongation of the foot (which is reduced to one toe), which has added another movable segment to the limb. The same is also true for many other cursors, including cats and ostriches. Having an additional segment not only increases the length of the lower limb relative to the upper one but also contributes to the speed of the stride. This increase in speed is because the speed of the foot is a summation of the speed of the individual segments of the limb.

When we think of animals in locomotion we naturally visualize the limb muscles as actively shortening, performing work to propel the animal forward. However, a recent study performed on turkeys running on a treadmill shows that this view may not be true (Roberts et al. 1997; also see Pennisi 1997). In this study a strain gauge was attached to the ossified tendon of the gastrocnemius muscle, and a device was implanted into the muscle to measure how much it shortened. When the turkey was running on the treadmill it was found that the muscle shortened by only a small amount while the foot was on the ground, and therefore performed very little work. Instead, the muscle generated a large propulsive force, and much of this force, surprisingly, was produced passively by the springlike stretching of the muscle. It was only when the leg was being swung forward in readiness for the next step that the muscle shortened significantly. The situation changed when the treadmill was tilted and the turkey had to run up an incline. Under these conditions the muscle did shorten significantly when the foot was on the ground, so considerably more work was being performed. The implication of this study is that skeletal muscles may have a different mechanical function during running than during other activities like jumping, flying, and swimming.

More on Transport Costs

The subject of transport costs has occupied the attention of a large number of biologists over the last few decades. A general picture has emerged of the relationship between the metabolic costs of terrestrial locomotion and body size. Many experimenters have put animals on treadmills, assessing their costs of transport by measuring the amounts of oxygen consumed during a given interval of time. One of the more unexpected findings is that the energy cost of transport, expressed in joules per kilogram per stride, at the same equivalent speed (as in the transition from trotting to galloping) is fairly constant over a wide range of animals, independent of body mass (Heglund and Taylor 1988). The energy cost at the trot/gallop transition is 5.3 J/kg/stride. This means that a buffalo and a warthog both have to expend about 5.3 joules per stride when they are running at the trot/gallop transition to transport each kilogram of their body mass. Although their cost per stride is the same, the warthog, being much smaller, takes more strides per second at this equivalent speed than the buffalo. It follows that the warthog's transport costs are relatively higher, simply because it takes more strides. It has also been suggested that smaller animals, with their more rapid strides, have to use faster muscle cells at a given speed than do larger animals. Since the faster contracting muscle cells cost more to operate than the slower ones, this exacerbates the situation. Small animals therefore have higher transport costs than large animals, which parallels their higher basal metabolic rates. Another way of looking at the lower transport costs of large animals is in light of the results of the experiments with the running turkeys. If muscles are generating force rather than performing work during locomotion, then the differences in costs are attributable to differences in the forces produced by the muscles. Large animals have longer duty factors than small ones, and their muscles therefore develop forces more slowly, which requires less energy (see discussion in Pennisi 1997).

The cost of transport during trotting and galloping generally increases with the stride frequency used to maintain a constant speed at the various equivalent speeds. For example, the transport costs at the trot/gallop transition is 5.3 J/kg/stride, compared with 7.5 J/kg/stride at the preferred galloping speed and 9.4 J/kg/stride at the maximum sustainable galloping speed (Heglund and Taylor 1988). This increase in transport cost may be related to the need to employ larger muscle forces as the mechanical advantage of the limb changes with increasing stride length.

My wife, who exercises by taking long, fast walks, claimed that she "burns off" just as many calories when walking as I do when I am jogging. Naturally, I disagreed, but we first had to be sure of our argument. Was her claim based upon our conducting our respective exercises over the same time, or over the same distance? If the former, I would cover about twice the distance, because I jog about twice as fast as my wife walks. Under these conditions she was happy to concede that I would use more energy. However, when the question became who uses more energy in a given distance, I had a real argument on my hands. The variation in physical factors, such as the difference in wind resistance (drag) between a brisk walk and a trot are of no consequence and can be dismissed. I therefore argued that my muscles have to generate greater forces to produce larger accelerations in my legs, in keeping with my higher stride frequency, but that did not convince her. However, when I got on the treadmill in the basement and conducted some simple experiments, I think I convinced her that I was right. This experiment will be explained later, but before turning to the practical, I want to conclude this section by saying something about size and speed.

Running Speed and Body Mass

We have seen that larger animals take longer, but fewer, strides. Conversely, small animals make up for their short stride lengths by taking more of them. This relationship between size and stride has led some specialists to argue that small animals should be able to run just as fast as large ones, and this argument can be supported by dimensional analysis. Dimensional analysis involves representing equations in terms of the basic dimensions of length (L), mass (M), and time (T). The speed (v) of an animal can be calculated by multiplying stride length (s) by stride frequency (f). This can be expressed by the equation, $v = s \times f$. So if an animal had a stride length of two meters and a stride frequency of three strides per second, its speed would be six meters per second. If we represent the terms on the right-hand side of the previous equation in the dimensions of L, s becomes L and f becomes $1/L$ because stride frequency is inversely proportional to length.

But when L is multiplied by $1/L$ the Ls cancel out. We are therefore left with an expression for speed, v, that does not contain the dimension of length, meaning that speed is independent of size. Although this conclusion is counterintuitive and contrary to our everyday experience, a number of studies have shown it to be true. In one of these studies the investigators clocked the running speeds of ten species of African ungulates (hoofed animals) by chasing them in a truck (Alexander, Langman, and Jayes 1977). The animals ranged in size from a 20 kg (44 lb) Thomson's gazelle to a giraffe weighing fifty times as much, but they all had about the same top speeds. Regardless of the large size range, this is a very small sample size and is restricted to only one group of animals (ungulates). Not surprisingly, a different picture emerged from a study of over one hundred different species of mammals (Garland 1983). This more comprehensive study showed that top running speeds did indeed increase with body size. But the largest mammals were not the fastest, and speeds peaked at body masses of about 120 kg (260 pounds). Some caution is needed when considering running speeds, though, because of all the difficulties of trying to record them. Even if you can persuade an animal to run alongside your truck while you clock its speed, the chances of its running in a straight line are remote, which affects the accuracy of the measurement. Furthermore, although the animal may be anxious to escape the truck and its annoying occupants, there is no guarantee that it is running at its top speed. Nevertheless, these records of running speeds are all we have to work with, and since they all suffer from similar measurement errors, they are likely to give us a good indication of the relative running speeds of a wide variety of mammals. They show that up to an optimum body size, speed increases with body mass.

PRACTICAL

Experiment 1   Determining your center of mass.

You are provided with a plank of wood that is as long as you are tall (a piece of two-by-four works quite well), a short length of dowelling (about 20 cm long and 2 cm in diameter), and some masking tape.

This experiment requires two people. Balance the plank of wood on the dowelling, like a teeter-totter (seesaw). Lie down on the plank so that the top of your head and your heels are about the same distance from either end. Press your arms against your sides. You will probably find that the head end of the plank touches the floor. Ask your partner to roll the dowelling toward the head end of the plank while you continue to lie on it. When your body balances, the dowelling will be at the level of your center of mass. Your partner marks the level by sticking a piece of masking tape to the outside of your clothing. As a rough guide, your center of mass is about one palm's width below the level of your navel.

Experiment 2   Finding how the stability of a quadruped is determined by the position of its center of mass.

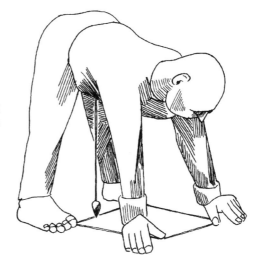

Figure 10.5. Adopting a quadrupedal stance for some simple experiments on balance.

You are provided with some rubber bands, a paper clip, some thread, and some masking tape. Link the rubber bands together to form a continuous loop that will stretch to about your height.

In this experiment you are going to emulate a quadruped. Tie the paper clip to the end of the thread. Using the masking tape, attach the other end to your abdomen, at the level of your center of mass. Adjust the length of the thread so that the paper clip hangs just below the level of your knees.

Remove your shoes and socks and adopt a quadrupedal stance, spreading your hands and feet so they form a rectangle (Fig. 10.5). Stretch the rubber loop between your hands and feet, holding it in place with your thumbs and big toes. The paper clip suspended from your center of mass should dangle in the middle of the rectangle formed by the rubber loop, almost touching the floor. (When you determined the position of your center of mass your hands were at your sides and your feet were touching. Your new quadrupedal pose has changed this a little, but the small shift in center of mass is not significant for these experiments.)

Leaving your right hand in place, release the rubber from your right thumb. The rubber now forms a triangle between your left hand and your two feet. The paper clip should fall just outside of the triangle. This means that your center of mass lies outside of the tripod formed by your left hand and two feet. What do you predict will happen if you lift your right hand from the floor?

You should have predicted that your body would be unstable, and you should have toppled toward the unsupported right side. The only way to prevent yourself from toppling is to redistribute your weight so that the position of your center of mass falls within the three supports again.

Repeat the experiment, but this time release the rubber from one of your feet. You will find that the position of your center of mass now falls within the three supports. You can predict from this that removing your foot from the floor will not cause you to topple. Try it.

Experiment 3   Examining how the stability of a biped is determined by the position of its center of mass, and how stability can be enhanced.

In addition to the thread and paper clip from the previous experiment, still attached to the level of your center of mass, you are provided with the plank, as used in the first experiment.

The masking tape attached to your person marks the *level* of your center of mass, but the actual point lies beneath the surface, somewhere between your abdomen and back. Bear this in mind in the following experiment where the dangling paper clip will lie somewhat anterior to the true position of your center of mass. Adjust the length of the thread so that the paper clip almost reaches your feet.

With your hands held behind your back, lean forward, bending slightly at the waist. Look at the paper clip, which should be dangling just above your feet. (If necessary, adjust the level of the thread.) Lean farther forward. Provided your center of mass lies within the support area of your feet, you will not topple forward. Test this by leaning progressively farther forward until you become unstable.

Kangaroos have long tails. Dinosaurs had long tails, too, and many of them were bipedal. The primary role of the tail of these bipedal animals was to stabilize the body. You can emulate a bipedal dinosaur by holding the plank of wood out behind you, like a tail. Keeping your "tail" well clear of the floor and roughly horizontal, try leaning forward again. How far can you lean this time without toppling?

You should find that you can lean much farther forward. This is because the plank has shifted your center of mass farther back, so you can lean farther forward while still keeping it above your feet.

Lean as far forward as you can, without toppling. With an assistant standing by to catch it, let go of the plank. You will find that you immediately topple forward as your center of mass extends beyond your feet.

Experiment 4   Investigating sexual dimorphism in the position of the center of mass.

Compare the position of your center of mass with that of a member of the opposite sex. You will find that it is relatively lower down in the female, which relates to sexual dimorphism in the pelvic region, associated with childbearing. The difference in position of the center of mass is the rationale behind the following party trick with a matchbox.

Kneel down in a praying posture, hands flat together, with your buttocks touching your heels. With your forearms on the floor in front of you, and your elbows touching your knees, hold a matchbox (or similar sized object) between the tips of your fingers and place it, end down, on the floor, as far away from your knees as possible. Clasping your hands behind your back, lean forward and try to touch the top edge of the matchbox with the tip of your nose. Women will be able to do this, but men will invariably topple forward.

Experiment 5   Measuring stride length at different gaits and speeds.

Your equipment needs for this simple experiment are determined by how you plan to record your footfalls. If you live in a climate like mine, you might do this experiment outdoors in winter and leave your footprints in the snow, in which case all you need is a tape measure. Alternatively, if you do not mind the puzzled reaction of your neighbors, you can do this on a dry sidewalk by wetting the soles of your shoes in a suitable container of water. If you are really fortunate, you might like to do this on a beach, marking your footprints in the damp sand. If you prefer the privacy of indoors, there are several ways of marking your footprints, depending on the floor covering. If there is a carpet, you can try making a heavy chalk mark on the sole of one of your shoes. This will leave a mark on the carpet, but there are limitations to how many strides you can take before the chalk mark left on the carpet becomes too faint. If the floor is bare, the best way is to soak a small piece of sponge in food dye and attach it to one shoe with a stout rubber band.

Once you have worked out how to mark your footfalls, walk for several meters at a leisurely pace. Now go back and measure your stride length. Remember that this is the distance between one footfall and the next footfall of the same foot. It does not matter what part of the foot you take as your mark, provided you always measure to the same point. For example, if you are measuring footprints in the snow, always measure to the tip of the imprint, or to the back of the heel. Repeat your stride-measuring experiment, but this time go for a brisk walk. You should make the last series at your fastest sprint.

The results of these experiments are very predictable. Your stride length for a brisk walk is longer than for a slow walk. During the run, you would have found an increase in stride length as you accelerated to maximum speed. You might be surprised at the length of your stride at a full sprint.

If you have a chance to compare stride lengths of people of different heights, you would find that tall people, predictably, have longer stride lengths than short people at the same equivalent speeds.

Experiment 6   Comparing the costs of transport during walking and running.

This experiment requires the use of a powered treadmill that gives a readout of speed and heart rate.

Assessing the cost of transport requires an apparatus for measuring the amount of oxygen consumed in a given time. However, you can make an approximation by using heart rate as an indicator of metabolic rate.

Set the speed for a brisk walking pace of about 5 km/hr (3 mph). Keep walking for about five minutes or so, to allow your heart rate to stabilize. Record this rate. Next, have another person measure the vertical displacement of your center of mass. This can be done simply by measuring the vertical movement of your head, either directly, or from your shadow cast on a wall. Also record your stride rate. It is not possible to measure your stride length, but you can get some idea of it from recording the distance between the footfalls of your left and right foot.

Keeping the speed setting constant, change gait to a jog. Keep up the jogging for several minutes, as before, to allow your heart rate to stabilize. Record this rate, together with your vertical displacement, stride rate, and an assessment of your stride length.

Repeat the entire sequence so that you have several episodes of running and walking.

When I did this experiment I found a satisfying relationship between the variables measured. When I changed from a walk to a run, my heart rate, vertical displacement, and stride rate all increased, while my stride length decreased. A decrease in stride length when changing from a walk to a run is contrary to the norm, but remember that your speed remains constant, and since your stride rate increases, you have to decrease your stride length. The additional physiological costs of increasing stride rate, causing more rapid muscle contractions, together with the additional mechanical costs of raising the center of mass through a greater distance, are reflected in the increased heart rate.

This experiment settled the argument I had with my wife about the energetic costs of walking and running – running a mile obviously uses more energy than walking a mile.

Experiment 7   Investigating the effects on running performance of changing the inertia of the legs.

You are provided with a set of ankle weights, as used by runners for training, some duct tape, and a stopwatch. Attach the weights around your waist, using the duct tape if necessary, and time yourself over a short dash of about 30 meters. Repeat with the weights attached to your ankles. Repeat with the weights attached as high up your leg as practical. You will probably have to use the duct tape to hold them in place. Alternatively, you may be able to effectively attach them to your thighs by inserting them into your pockets.

You should have found that your running became progressively more strenuous and probably slower as you increased the inertia of your legs (weights lower down). This is because your muscles had to generate greater forces and therefore consume more energy. Transport costs therefore increase with leg inertia.

Experiment 8   Assessing the locomotory costs associated with displacing the center of mass during walking and running.

You are provided with a piece of colored chalk, a meter rule (or equivalent), a blank wall, and a stopwatch.

You can obtain a reasonable assessment of the displacement of the center of mass during locomotion by analyzing the wave form generated by a fixed point on the body as it moves past a fixed vertical surface. To record this wave form, hold a piece of chalk against a wall, at a fixed height, and write a line as you walk. You must take care not to move the position of your hand during the procedure. If there is no suitable wall, you can tape a roll of wallpaper to an indoor wall, reverse side out, using masking tape. You can try using colored pencils or magic markers instead of chalk.

You should obtain a sinusoidal line. What is the relationship between cycles and strides? Your vertical displacement can easily be measured from the chalk tracing. Simply line up the ruler with the bottom of two consecutive waves and measure the maximum amplitude of the crest. Measure several wave crests to obtain a series of displacements, taking an average. From your results, calculate the amount of work (joules) done for each cycle (see worked examples at end of chapter).

How much energy have your muscles had to generate to perform this amount of work? Remember that muscles are only about 20 percent efficient, therefore they have to generate about five times more energy than that required for the work to be performed. So only about one-fifth of the energy generated by the muscles appears as work; the remaining four-fifths appears as heat. Estimate the amount of energy your muscles had to generate to perform one cycle, assuming an efficiency of 20 percent. Now estimate the total amount of energy generated by your muscles during your short walk. Express this amount in terms of joules. For comparison you can also express your results in calories. Remember that there are 4.2 J per calorie. The calories referred to in cookbooks are kilocalories (kcal), or large calories, written with an uppercase C to distinguish them from calories (the amount of heat required to raise the temperature of 1 gram of water through 1 C°).[3] Partly because of the confusion associated with the term, calorie is not part of SI (the International System of Units).

Wipe off your chalk marks on the wall as you finish with them, using a damp cloth. Alternatively, alter the vertical position of the wallpaper or use a fresh length.

Repeat the experiment, but this time use the stopwatch to determine how long it takes to complete a series of cycles. From your results, calculate the amount of power (watts) needed to elevate your center of mass during this time interval. Estimate the total power generated by your muscles during your walk, assuming 20 percent efficiency of muscles.

Repeat the first part of the experiment while running. How does your vertical displacement compare with that obtained for a walk? Calculate the energy required during a series of elevations, as you did for the walk. Also estimate the total power generated by your muscles during your short run.

What assumptions are you making in these various calculations?

Worked examples:

Suppose your vertical displacement in the first part of experiment 8 was 3.5 cm, and that your mass was 77.3 kg. Given the acceleration of gravity, g, is 9.81 $m/s/s$, the force required to produce this displacement is:

Mass × acceleration = $77.3 \times 9.81 = 758.31 N$

The work involved is given by:

Force × distance = $758.3 \times 0.035 = 26.54 J$

Assuming a muscle efficiency of 20 percent, the actual amount of work your muscles had to perform is:

$5 \times 26.54 = 132.7J = 132.7/4.2 = 31.59$ calories

Suppose you completed six cycles of displacement in 5.8 seconds. The work performed is $26.54 \times 6 = 159.24J$. The rate of doing work is $159.24/5.8 = 27.45W$ (1 watt is 1 joule per second). Assuming a muscle efficiency of 20 percent, the total power generated by your muscles is $27.45 \times 5 = 137.28W$.

Fluid Flow

Most people think of fluids as something that comes out of bottles, but the term *fluid* applies to liquids and gases. The important feature shared by liquids and gases is the mobility of their molecules. Fluids are therefore free to flow, and the nature of this movement has been the subject of intensive study, resulting in a vast literature. The reason why such a seemingly esoteric topic as fluid flow has attracted so much attention is because it has so many practical applications. These include the design of aircraft, ships, fuel-efficient cars, gas pipelines, dams, fuel injectors, turbines, pumps, windmills, propellers, and fans. The knowledge of fluid mechanics has also permeated our leisure world, shaping a variety of things from the wheels of racing bikes to the way ski jumpers hold themselves, and their skis, during flight. It is only through an understanding of fluid mechanics that we can begin to comprehend how birds fly and fishes swim, or make sense of features like the ragged wing tips of crows or the crescentic tails of mackerels. We will therefore devote this entire chapter to the physical aspects of fluid flow, starting with the subject of drag.

Drag: The Resistance to Forward Motion

We are surrounded by a fluid that is not very dense, so we are usually not aware of drag, or wind resistance, unless we are traveling especially fast. If I walk down the corridor I cannot feel the impact of the air against my face. However, if I am late for an appointment and need to run, I can feel the air as it streams past me. Still, the drag force is negligible, and certainly not enough to slow me down perceptibly. If I were riding a bike downhill, the drag force would be sufficient to slow me down a little, but I could reduce this effect by crouching low over the handlebars. In doing so I would be reducing my *frontal area,* which is the cross-sectional area of the moving body, measured perpendicular to the fluid flow. The drag force on a motorcyclist traveling at high speed is considerable, and if he were foolish enough to stand up on the foot pedals, thereby increasing his frontal area, he would likely be swept clean off his machine. One of the best ways to experience drag is to put your hand out of the window of a car. If you hold your hand parallel to the flow, by pointing your palm toward the road, the frontal area is minimal and the drag force is accordingly small. However, if you now rotate your hand so that the palm faces the flow, the frontal area is maximal and the drag force is much greater. Indeed, if the car is traveling fast, your arm is likely to be swept back by the large drag force.

You have to be traveling fairly fast in air to appreciate drag forces, but if you are traveling in water, which is over 800 times more dense than air, the drag is appreciable, even at fairly slow speeds. Trailing a hand over the side of a canoe or running it through the water in a bathtub makes you realize how large drag forces can be. Anything that can be done to reduce drag forces, whether on a ship's hull, a turbine's blade, or an airliner's fuselage, reduces energy losses, hence costs, and is therefore at a premium. The same is true within the animate world, and since fishes and birds face the same fluid forces as ships and planes, we see similar resolutions to the problem of reducing drag. One universal mechanism for reducing drag is the streamline, but before looking at how it works, we need to consider how fluids flow and what forces they exert on a body.

Laminar and Turbulent Flow

If you turn on a tap gently, the water can be made to flow smoothly, like glass, but if you turn it on fully, the stream breaks up into a disorganized torrent. A similar contrast between smooth and rough flow can be seen when you take a boat out on a smooth lake. When you cruise slowly under minimum power, the water flows smoothly and gently down the sides of the boat, each layer slipping past the next without any swirling movements. Such flow is described as *laminar,* and the drag force exerted on the boat by the water is minimal. But when the throttle is opened up and the boat surges forward, the water breaks up into a swirling maelstrom. Such a rough and disorganized flow of fluid is described as being *turbulent,* and the drag force exerted on the boat is considerably increased over the laminar situation.

Most bodies, animate or inanimate, generate some swirling movements when they move through a fluid, unless they are very small or are moving very slowly. The energy for setting the fluid particles in motion comes from the body itself, so the swirling movements create much of the drag force. If a body could move through a fluid without causing turbulence, the drag force would be greatly reduced.

A useful way of examining drag forces is to visualize what happens when a thin plate is moved through a fluid. Actually, most laboratory experiments on fluid flow have the object stationary and the fluid in motion, as in experiments conducted in wind tunnels. This is purely a matter of convenience, and it makes no difference to the end results. All that matters is that the flow of fluid is *relative* to the object being tested.

The most turbulence is achieved when you place the plate perpendicular to the flow's direction (Fig. 11.1A). You can think of the fluid as being made

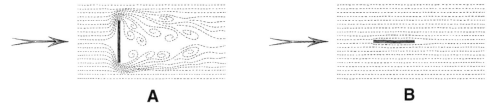

Figure 11.1. When a thin plate is placed perpendicular to the fluid flow (A), the drag forces are maximal. When it is placed parallel to the flow (B), the drag forces are minimal.

up of numerous small particles, some of which strike the plate's leading surface, opposing forward motion head-on. Many more are diverted before they can reach the plate; some move inward and create swirling eddies called *vortices*. Extensive vortices can be seen downstream of the plate. Energy is required to move each fluid particle around the plate and also to accelerate the particles in new directions. How much energy this movement requires depends on the particle's inertia – the tendency for it to stay where it is – which is proportional to its mass and therefore to the fluid's density (mass per unit volume). This component of drag is referred to as *pressure drag* or *form drag*. Almost all the resistance experienced by the perpendicular plate is due to pressure drag.

A second drag component, called *friction drag* (or skin friction), acts on the surface of the body, parallel to the direction of flow. Friction drag is attributable to the viscosity of the fluid, a concept we encountered earlier in the discussion of thick-film lubrication (Chapter 8). When the plate is placed perpendicular to the fluid flow, friction drag is negligible. This is because hardly any surface area is parallel to the free stream, so almost all resistance to forward motion is due to pressure drag. The reverse holds true when the plate is placed parallel to the flow (Fig. 11.1B). The reason is that the fluid flow is now almost entirely laminar, and most of the resistance to forward motion is attributable to friction drag.

The total drag acting on a body is a combination of pressure drag and friction drag, attributable to the fluid's density and viscosity, respectively. The sizes of these two forces depend on the size of the body and the speed of the flow. A large and a small body therefore experience different forces even though they might be moving in the same fluid at the same speed. The relationship between these variables is expressed by a ratio called the *Reynolds number* after Osborne Reynolds, a British engineer and physicist, abbreviated Re:

$$\text{Re} = \frac{\text{length} \times \text{speed} \times \text{density}}{\text{viscosity}}$$

where density and viscosity refer to the fluid, and speed and length refer to the body moving through it. Whereas the length term is usually taken to be body length for swimmers, it is more usual to use the width of the wing (called the chord) for fliers. When calculating a Reynolds number, you must measure all of the variables in the same units, and because they cancel out in the equation, the Reynolds number has no units. In metric units, the density of water is 1000 kg/m^3 and the viscosity of water (at 10° C) is 1.304 × 10^{-3} Ns/m^2 (the units are impossibly obscure – newton-second per meter squared or pascal-seconds, PaS). Suppose we wanted to calculate the Reynolds number of an 18 m (59 foot) long blue whale moving at 2 m per second (7.2 km/h or 4.5 mph). The Reynolds number of the whale at this speed is:

$$\frac{18 \times 2 \times 1000}{1.304 \times 10^{-3}} = 27.61 \times 10^6$$

which is quite high. How about a smaller animal, say a 15 mm tadpole swimming in an aquarium? Small animals move more slowly than large animals, and a reasonable speed for a tadpole might be about 20 mm per second.

$$\text{Re} = \frac{0.015 \times 0.020 \times 1000}{1.34 \times 10^{-3}} = 224$$

which is quite low. Very small bodies, like dust particles in air and small planktonic organisms in water, move at Reynolds numbers of less than one. Movements at such low Reynolds numbers are dominated by viscosity forces, and the greatest drag component is therefore friction drag. This is irrespective of whether the fluid has a high or a low viscosity. Air, for example, has a low viscosity, but dust particles illuminated by a shaft of light in a darkened room appear to be floating in syrup. At high Reynolds numbers the situation is reversed: pressure drag is the main force, and viscosity (hence friction drag) is of secondary importance. But friction drag is always present and can never be ignored.

I teach a marine biology course most summers, and I do a simple experiment that demonstrates how large the frictional component of drag can be. When we are out on a boat, cruising along at about 4 knots (7.4 km/hr or 4.6 mph), we let out about 100 yards (92 m) of thin line, which trails through the water at a high Reynolds number (Re = 1.5 × 10^8). The frontal area of the line is relatively small, so pressure drag is probably negligible.[1] But the surface area parallel to the flow is relatively large, and most of the drag is due to friction. I ask the students to think about the friction drag when they grab hold of the line, and they are always amazed at how hard they have to pull to overcome the drag force. (Much of the friction drag is probably due to the turbulence in the boundary layer.)

In addition to the drag forces changing with the Reynolds number, the thickness of the boundary layer also changes. This inverse relationship between Re and the thickness of the boundary layer is evident every time you drive a car in the rain. At low speeds the Reynolds number is low and the boundary layer is correspondingly thick. The beads of water that form on the windows lie within the boundary layer and therefore are not exposed to the full force of the air's free flow; they are not displaced by the air flowing over the car. As the car accelerates, the boundary layer decreases in thickness until it is thinner than the beads of water. At this point the beads that project beyond the boundary layer are exposed to the full force of the free airflow and are swept back by it. The largest droplets are the first to move, because they project farthest. You have to drive very fast before the smallest ones are swept away.

There is no clear cutoff point between high and low Reynolds numbers, and many organisms (and vehicles) operate over a range of values. Reynolds numbers in the thousands are high, those in the hundreds and tens are fairly low, and those less than one are definitely low. The following approximations will give you some idea of how they vary among different fliers and swimmers: jet aircraft and nuclear submarines 10^8; large whales 10^7; sharks 10^6; albatrosses 16,000; hummingbirds 8,000; water beetles 5,000; dragonflies 2,000; guppies 300; protozoa less than 1; and bacteria 10^{-4} and less.

Movements at low and at high Reynolds numbers are so fundamentally different that organisms living within the two regimes face quite distinct suites of problems. As a consequence, these organisms usually possess different structural adaptations. The streamlined shape, for example, is a way of reducing pressure drag. Since pressure drag is a minor component of drag at low Reynolds numbers, the streamlined shape is primarily confined to organisms and devices that operate at high Reynolds numbers.

The Streamlined Body and How It Reduces Drag

If a circular plate is placed in a fluid, perpendicular to the direction of flow, the vortices generated – and hence the pressure drag – is maximal. Replacing the disc with a sphere of the same diameter improves the situation, approximately halving the drag (Kermode 1972). The flow is essentially laminar around the front or *leading surface* of the sphere, and it continues like that beyond the widest part of the sphere, called the *shoulder,* toward the back or *trailing surface*. Behind the sphere the fluid flow becomes turbulent, and numerous vortices are generated (Fig. 11.2A).

If the leading surface of the sphere is drawn out into a rounded point and the trailing surface is gently tapered, the vortices can be essentially eliminated and the resistance is reduced to about 5 percent that of the perpendicular disc (Fig. 11.2B). This torpedo shape is called a *streamlined body*. The optimum streamlined shape for minimizing total drag has the shoulder placed between about one-third and one-half of the way back from the front of the body.

A useful way of expressing the relative slenderness of a streamlined body is by the ratio of the length of the body to its diameter. This is called the *fineness ratio,* or slenderness ratio. Increasing the fineness ratio decreases the pressure

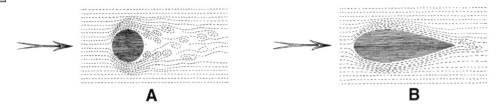

Figure 11.2. When a sphere (A) is placed in a moving fluid, the flow is essentially laminar around the leading surface but becomes turbulent beyond the shoulder. Streamlined bodies (B) generate little turbulence and therefore a smaller wake.

drag, but since it also increases the surface area, the friction drag is correspondingly increased. There is an optimum fineness ratio at which the total drag is minimal, and this has been found to correspond to a value of between three and seven for bodies moving in water (Weihs and Webb 1983).

The major drag component of a streamlined body is attributed to friction, which involves the boundary layer. The fluid flow in the boundary layer, like that in the free-flowing fluid beyond it, can be either laminar or turbulent. If laminar flow could be maintained in the boundary layer, friction drag would be reduced to about 10 percent of what it would be if the flow were turbulent (Kermode 1972). This ideal is probably never achieved, but it can be approached.

The flow in the boundary layer usually starts off being laminar over the first part of the leading surface. But a transition region is eventually reached, somewhere along the trailing surface, beyond which the flow becomes turbulent. The position of the transition region moves forward as the body moves with increasing speed, so as the body moves faster, more of the boundary layer becomes turbulent, thereby increasing the friction drag. As surface roughness is one of the factors that causes the boundary layer to become turbulent, having a smooth body is one effective way of reducing friction drag. Competitive sailors take great pains to make the hulls of their sailboats as smooth as possible, thereby promoting laminar flow.

Not only does the boundary layer become turbulent with increasing speed, but it also begins to separate from the body's surface. Separation begins at trailing surfaces and moves forward as the body gains speed. The separation in the fluid generates large vortices that reduce, then eliminate, the beneficial effect of the streamlined shape. Separation, like turbulence, can be caused by rough surfaces, which is another reason for having a smooth body. However, there are some circumstances, discussed later, in which roughness can actually be more beneficial than a perfectly smooth surface.

Just how smooth the body's surface must be to promote laminar flow and delay separation depends on the thickness of the boundary layer, which is determined by the Reynolds number. Modern aircraft, for example, fly at such high Reynolds numbers that the boundary layer is only about as thick as a paper clip. Even small projections extend well into the boundary layer, causing turbulence and promoting separation, which is why the rivets that secure the skin to the airframe have to be flush with the surface. Take a look next time you board an aircraft. Separation is especially critical on wing surfaces because it reduces the lifting capacity. This explains why ice buildup on wings can have such disastrous consequences.

Although smoothness and laminar flow are usually advantageous, sometimes roughness and turbulence are more beneficial. At high Reynolds numbers, for example, a turbulent fluid may actually remain in contact with a surface for longer than it would if its flow were laminar. This is why golf balls have dimples. The turbulence the dimples generate in the boundary layer delays separation so that the air stays in contact with the surface longer, actually reaching the trailing surface before breaking away. Consequently the width of the wake, and hence the drag, is much smaller. It has been estimated that a swing that would drive a regular golf ball for 230 yards (212 m) would drive a smooth one for only about 50 yards (46 m; Shapiro 1961). It is not difficult to visualize the evolution of the dimpled golf ball from smooth-surfaced beginnings. As the smooth surface of the early golf balls became cut by the impact from the club, it must have been noticed that they tended to go farther. Some enterprising golfer no doubt exploited this, leading to the invention of the dimpled ball. A friend tells me of a small golf course close to the sea where smooth balls are used in order to reduce the chances of golfers losing them in the sea. The same principle as the golf ball's dimples is at work in certain parts of aircraft. But instead of using dimples, the aircraft sets the airflow into turbulence with small rectangular projections called vortex generators. You can see these projections on the wings of the Boeing 757, for example, toward the leading edge.

The idea of deliberately causing air to become turbulent to *reduce* drag is counterintuitive, so how should it be rationalized? A. C. Kermode (1972), mindful of the challenge of explanation, asks us to think of the boundary layer as becoming sluggish as it passes along the surface of an airfoil. And as the boundary layer becomes slower, it becomes more prone to separation. Stirring up the boundary layer causes some mixing with the faster moving free stream, imparting additional energy. This mixing helps the boundary layer to travel farther along the surface of the airfoil before slowing down sufficiently to separate. The additional drag caused by setting the boundary layer into turbulence is more than compensated for by the reduction in drag due to the decrease in the size of the wake.

Drag, Speed, and Size

The relationship between drag and the speed and size of a body depends upon the Reynolds number, and also upon what expression of size is being assessed. At low Reynolds numbers, drag is proportional to the velocity, but at high Reynolds numbers it is proportional to the square of the velocity (Shapiro 1961). This relationship helps explain why large pelagic animals like swordfishes, which move at high Reynolds numbers and appear to be adapted for high speeds, spend most of their time cruising fairly slowly (Chapter 9). Two measures of size are important in fluid mechanics, namely, frontal area and length. At low Reynolds numbers, frontal area has a small effect on drag compared with length, whereas the reverse is true at high Reynolds numbers. We will investigate most of these points during the practical.

Inclined Plates, Angles of Attack, and Aspect Ratios

So far we have considered resistance to surfaces that are either parallel or perpendicular to the flow of a fluid. But what of surfaces that are between these two extremes – what forces are experienced by objects that meet the flow, or move through a fluid, at an angle? When a plate (a flat surface, also referred to as a plane) is inclined at a small angle to a moving fluid such that the leading edge is higher than the trailing edge, it experiences an upthrust or lift force. This is because the plate deflects fluid away from its lower surface, and the force applied to the fluid has an equal and opposite reaction on the plate. You can readily demonstrate this phenomenon by making your hand into an inclined plate and holding it out the window of a moving car. If you tilt your hand from the horizontal position, you will feel it being pushed upward by the air.

If the angle of an inclined plate to the fluid flow, called the *angle of attack,* is gradually increased, the magnitude of the lift force increases – but so, too, does the drag force. The lift force does not keep on increasing, however; it reaches a maximum at an angle of attack of 10–20°, after which it begins to decrease. However, the drag force increases all the time, reaching a maximum when the angle of attack reaches 90°. If the angle of attack is increased beyond 90°, that is, if the plate is inclined downward, the plate experiences a downthrust instead of a lift.

The rudder of a ship is a vertical inclined plate that can be inclined to one side or the other. When the rudder is moved so that its trailing edge moves to the right, the thrust is to the left. This force pushes the stern to the left, which in turn swings the vessel around and points the bow to the right.

The effectiveness of an inclined plate can be expressed by the ratio of the lift force to the drag force, called the *lift-to-drag ratio.* The lift force increases with increased area, but so does the drag force, so we cannot improve the lift-to-drag ratio simply by increasing the size of the plate. However, a long, narrow plate, with its long axis at right angles to the flow, will have a better lift-to-drag ratio than a square plate. Thus, if there were two plates of equal area, one square and the other long and narrow, the second plate would generate the greater lift for a given drag force. The relative narrowness of an inclined plate is expressed by the *aspect ratio,* which is the ratio of the length to the width. A plate 10 units long and 10 units wide has an aspect ratio of one, whereas a plate that is 20 units long and 5 wide has an aspect ratio of four. One of the reasons why inclined plates (and wings) with high aspect ratios have a higher lift-to-drag ratio is that they generate relatively small vortices at the tips, a subject dealt with in the next chapter. The lift-to-drag ratio of a plate is also increased by its having a stream-lined profile, because at a given high Reynolds number, a streamlined body experiences lower drag forces than one with a rectangular profile.

PRACTICAL

This is an extended practical section. It will explore more thoroughly several of the topics that were dealt with only briefly in the text. It is quite likely that you may not have access to a wind tunnel (Part 1). If that is the case, you should still work your way through the material because you will find some important new information here.

PART 1

As this segment involves the use of a wind tunnel, I will begin by describing how this equipment works and how it can be used for measuring horizontal forces (drag) and vertical forces (lift). The system used here is an open wind tunnel, so called because the models to be tested are placed in the open, in front of the outlet from the machine (Fig. 11.3A). The large wind tunnels used in the aircraft industry, in contrast, are usually of the closed variety, where the test models are placed inside the equipment. The wind tunnels we use at the University of Toronto were manufactured by a British laboratory supplier some years ago. As you may not have such equipment available to you, I will give sample data for all of the experiments conducted. By reading through the descriptions of the experiments and working your way through the data, you should get a good idea of how the system works and of the underlying principles being investigated.

The wind tunnel is a simple tube, constricted at the outlet and fitted with vertical and horizontal baffles, downwind of the fan, to smooth the air-flow. The speed of the fan is controlled by a rheostat (variable resistor), which is marked off in settings from zero to nine. The airspeed has been measured for each of these settings and is given in meters per second. A value for the velocity of the air measured at right angles to the main flow is also given, in meters per second. The latter is a measure of the degree of turbulence. These values are small compared with the velocity of the air parallel to the longitudinal axis of the wind tunnel, meaning that the airflow is essentially laminar.

The airflow is not especially fast, nor are the models particularly large. However, at the fastest setting (14 m/s; approximately 50 km/h or 30 mph) and with the larger models, the Reynolds numbers become quite high. For example, at the fastest setting, the Re for a 12 cm diameter sphere is 1.15×10^5. At high Reynolds numbers, recall, friction drag becomes relatively unimportant compared to pressure drag.

The balance to which the models are attached can be set up to measure

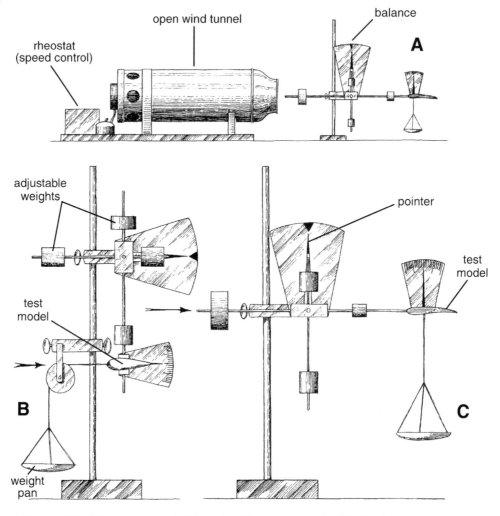

Figure 11.3. Using an open wind tunnel and balance apparatus (A). The balance can be set up to measure drag (B) or lift (C).

horizontal forces or vertical ones. In the first case, used for measuring drag, weights are added to a scale pan to counteract the horizontal drag force generated by the attached model (Fig. 11.3B). By adding weights to the pan so that the horizontal pointer stays on its starting zero mark, the model is kept in its original position. The force represented by the weights in the pan is therefore equal to the drag force tending to move the model from its original position.

Similarly, when the balance is set up to measure lift, the pointer is arranged vertically, and the upthrust generated by the airfoil model is counteracted by the weights added to the pan (Fig. 11.3C). The force represented by the weights therefore equals the lift force, tending to elevate the model and move the pointer off the zero mark.

In both cases the adjustable weights attached to the horizontal member of the apparatus are used to zero the pointer at the start of each experiment. The attached weights also have a damping effect on the balance, smoothing out excessive vibrations of the models in the airflow.

Experiment 1 Determining the drag forces generated by a disc oriented perpendicular and parallel to the airflow.

Set up the apparatus as shown in Figure 11.3B with the pointer horizontal. To do so, simply adjust the position of the weights. It is very important that you set up the weight pan and pulley correctly, otherwise your results will be meaningless. The model to be tested will be fitted to the lower end of the vertical beam. Set the pulley up on the same level as the support for the model such that the string, which passes from the weight pan and over the pulley, is *horizontal*. Also make sure that the pulley is vertically in line with the attachment hook (check the alignment from the side).

Can you see why it is important to have the string horizontal? Consider the moment arm of the force due to the weight pan and its weights (moment arm, remember, is the distance between the fulcrum and the force, measured at right angles to the line of action of the force). The moment arm here is the distance from the fulcrum to the point of attachment of the string, because the string is horizontal. When the model is pushed back by the drag force, lifting the weight pan, the moment arm barely changes (it is slightly reduced). If you had the pulley higher than the attachment hook, the moment arm of the weight pan force would be considerably reduced, and you would have to add more weights to compensate. (Try the effect of raising the pulley – you will find that weights have to be added to the pan to return the pointer to the horizontal position.)

Before you begin, you have two important checks to make:

1. Ensure that the pointer is horizontal (this can be checked with a spirit level) and lines up with a reference mark.
2. Make sure that the pointer swings freely and returns to the horizontal mark.

Now attach the disc to the apparatus so that it is perpendicular to the direction of the airflow. Adjust the position of the model so that it is in the center of the air outlet. Also make sure that the object is orthogonal (at right angles) to the outlet.

WARNING: If the center of mass of a model under test does not lie along the axis of the vertical beam, it will throw the latter off the vertical. If you do not believe this, attach the streamlined model to the beam and adjust the weights until the pointer lines up with the reference mark. Now rotate the model through 90° and see how the pointer moves off the mark. It is therefore important to check that the pointer is lined up with the horizontal scale each time you change a model or its orientation.

With the rheostat at the lowest setting, turn on the motor. Watch the pointer move upward as the drag upon the disc pushes it away from the fan. The motor takes a little time to reach its maximum speed at any particular setting of the rheostat, so wait for about half a minute to allow it to stabilize before taking any measurements.

Add weights to the pan until the pointer returns to zero (approximately 2.0 g at this air speed). The balance may tend to stick in one position, so give the weight pan a nudge when you think you have reached the zero position, just to make sure it is free.

Increase the speed of the airflow to the next speed setting. Add more weights to zero the scale again. You are working to an accuracy of only 0.5 g, so do not worry if the pointer does not come exactly to zero. Repeat the procedure for the remaining speed settings. Remember to nudge the pan each time you think you are at the balance point, to make sure that the balance has not stuck.

Repeat with the disc parallel to the airflow.

Sample Observations:

Airspeed (m/s)	Disc perpendicular (drag; g)	Disc parallel (drag; g)
3.9	2.0	0.5
4.5	2.7	0.5
5.1	3.5	1.0
6.0	4.2	1.0
7.0	5.7	1.5
8.0	7.5	1.5
10.2	10.0	2.0
12.0	13.5	3.0
14.4	18.5	4.0

From your results you can see the following:

1. Drag increases with increasing fluid velocity.
2. The disc generates considerably more drag when placed perpendicular than when placed parallel to the airflow.

Experiment 2 Examining the relationship between drag and velocity for a disc, a sphere, and a streamlined body, all of the same frontal area.

Set the apparatus up as in the previous experiment and measure the drag force on each of the three models in turn at increasing airspeeds, starting with the disc. Make sure that the disc is perpendicular to the airflow and that the long axis of the streamlined body is parallel to the flow. Tabulate your results.

Sample Observations:

Airspeed (m/s)	Disc (drag; g)	Sphere (drag; g)	Streamlined body (drag; g)
3.9	1.0	1.0	0.5
4.5	2.0	1.5	1.0
5.0	3.0	2.0	1.0

Airspeed (m/s)	Disc (drag; g)	Sphere (drag; g)	Streamlined body (drag; g)
6.0	4.0	2.5	1.0
7.0	5.0	3.5	1.0
8.0	7.0	4.0	1.5
10.2	9.0	6.0	2.0
12.0	14.0	7.5	3.0
14.4	20.0	10.5	4.0

From your results you can see the following:

1. Drag increases with increasing fluid velocity.
2. A sphere experiences less drag than a disc of similar frontal area, and a streamlined body experiences even less.

Plot graphs of drag, in grams (y axis), against airspeed, in meters per second (x axis), for the disc, sphere and streamlined model. Do your graphs look as if they may be exponential? ($y = bx^k$ where y = drag, x = airspeed, b = a constant, k = exponent.) Remembering how difficult it is to interpret exponential graphs, plot them again, this time using logarithmic data. The equation of the graph is now $\log y = \log b + k \log x$, which of course is a linear function. Measure the gradient of each graph ($dy/dx = k$), which gives you the value of the exponent, k.

I obtained three quite different exponential graphs for my raw data (Fig. 11.4A). Logarithmic data gave three straight lines, with gradients of about 2 for

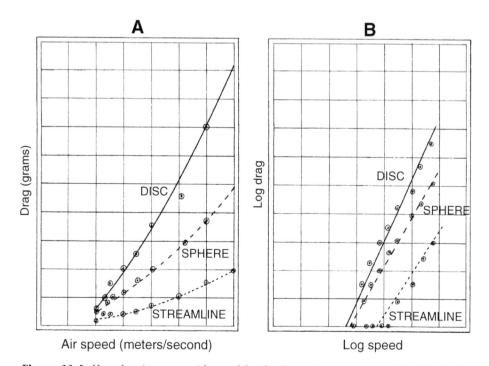

Figure 11.4. How drag increases with speed for the disc, sphere, and streamlined body: (A) raw data; (B) logarithmic data.

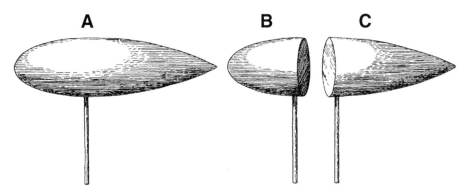

Figure 11.5. The complete streamlined body (A), the anterior segment (B), and the posterior segment (C).

the disc and sphere (approximately 2.2 and 1.8, respectively) and about 1.4 for the streamlined body (Fig. 11.4B). This experiment shows that drag is approximately proportional to the square of the airspeed for the disc and the sphere, but closer to unity for the streamlined body. The reason for this is that most of the drag generated by the disc and sphere is attributed to pressure drag, which is determined by the inertia of the fluid, whereas most of the drag generated by the streamlined body is attributed to friction drag, which is determined by viscosity. In situations dominated by inertial forces, drag is proportional to the square of the velocity, whereas drag varies with the velocity in viscosity-dominated situations where the flow is predominantly laminar.

Experiment 3 Analyzing the drag-reducing attributes of the streamlined body at a high Reynolds number.

You are provided with three solid models: a streamlined body; the anterior portion of the same body; and the posterior portion (Fig. 11.5). You will be testing each of the three models in turn, first with its tapered surface pointing toward the airflow, then the other way (Fig. 11.6). As the streamlined body is a high Reynolds number device, you will do all the testing at the highest speed setting. Tabulate your results.

Your results probably showed that the anterior and posterior sections of the streamlined body gave approximately the same drag forces when they were ori-

Direction of airflow

Figure 11.6. The various orientations of the streamlined shapes in the airflow.

Direction of airflow

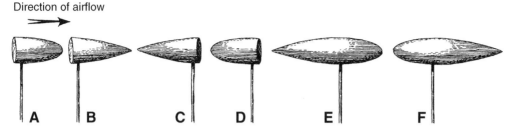

Figure 11.7. The various streamlined shapes, arranged in probable order of decreasing drag forces.

ented in the same way. In both cases the drag force is approximately twice as high when the flat surface faces the airflow. The probable reason for this is that the flat face generates a considerable amount of turbulence, producing a wide wake. The wake is much narrower when the rounded or pointed surfaces face forward. If you arranged the partially streamlined models in order of decreasing drag, you would probably find they corresponded to that depicted in Figure 11.7. The likely explanations for these results are that model B probably generates a marginally smaller wake than model A because of its more gentle taper. Models C and D probably generate wakes of similar widths, but because model C has a larger surface area, its friction drag is probably greater than that of model D.

When the streamlined model is oriented correctly (Fig. 11.7F), the drag force is lower than in any other model. The rounded front surface allows the fluid to flow over it without generating a great deal of turbulence, therefore the drag is small. Because the gradient of the posterior surface is gentle, the airflow maintains contact with most, or all, of the trailing surface, thus minimizing the size of the wake. When the model faces the other way (Fig. 11.7E), the pointed anterior surface cuts through the air well – probably much better than the rounded surface did before – and may produce little or no turbulence. However, the air does not maintain contact over much of the posterior surface because of the steeper gradient, hence the wake is wider than it would be if the streamlined model were placed the other way around.

Notice that the drag on the entire streamlined model, when oriented with its sharp end facing the airflow (Fig. 11.7E), is less than when its rounded surface is truncated (Fig. 11.7C). This fact has much relevance to the design of ships. Most boats are built with a sharply pointed bow for cleaving through the water, thereby reducing the size of the wake, but sometimes insufficient attention is paid to the stern. This is especially so for pleasure craft, which are often given a truncated stern. If this squared-off shape is continued below the water line, it adds considerably to the drag on the hull.

Remove the streamlined model and replace it with the rod provided. Balance the beam, then set the rheostat to no. 9. You will find that the drag force generated by the rod is about the same as that generated by the streamlined body, which is remarkable when you consider that the diameter of the rod is only about one-ninth that of the streamlined model. The wire rigging used for early biplanes generated considerable drag forces. Aircraft designers came to recognize this in the end and replaced wires with streamlined wooden struts. A strut had to be almost ten times thicker than a wire before it generated an equivalent drag force.

Experiment 4 Examining the effects of roughness on the performance of the streamlined body, at high Reynolds numbers.

Set up the streamlined model in the apparatus, as in the previous experiment, and measure the drag at the maximum airspeed. Record your results. Now add a ring of small (about 4 mm in diameter) plasticine bumps around the circumference of the leading surface of the model and measure the drag again. Add a second circlet of bumps to the trailing surface and record the new drag. You may like to experiment with different numbers of circlets and different positions.

You will find that adding small plasticine protuberances increases the drag, because the projections cause laminar flow to become turbulent, thereby increasing the size of the wake. I found that adding one ring of bumps to the leading surface increased the drag from 4 g to 5 g. Adding a second circlet to the trailing surface increased the drag to 7 g. However, when I added a third circlet, just posterior to the shoulder region, the drag dropped marginally, to 6.5 g. A possible explanation for this drop is that with the first two circlets in place, the air was tending to separate from the shoulder region. Adding the third set stirred up the air, causing it to become turbulent and thereby adhere more closely to the surface than had the previously laminar airflow. Five circlets of bumps increased the drag to 9 g.

Experiment 5 Investigating the relationship between drag and disc diameter, at high Reynolds numbers.

You are provided with a set of discs of increasing diameter. With the wind tunnel set at maximum airspeed, measure the drag for each of the discs, oriented perpendicular to the airflow. Tabulate your results.

Sample Observations (airspeed 14.4 m/s):

Diameter of disc (cm)	Drag (g)
3.2	13.0
4.3	24.0
5.0	30.5
6.1	43.0
7.4	58.0
9.3	82.0
14.4	165.0

Plot a logarithmic graph of drag (y axis) against diameter, and measure the slope. I obtained a value of about 1.7, which is sufficiently close to 2 to suggest a relationship between drag and (frontal) area, as predicted.

Experiment 6 Examining the relationship between drag and length, at high Reynolds numbers.

In assessing the relationship between drag and length, we have an acute restriction on the choice of shapes for the models. We cannot use spheres because the length of a sphere is also its diameter. If streamlined models were used, we would run into problems because we would have to keep the diameter constant. The longest model would therefore be long and thin, the shortest one stubby, so they would vary in shape as well as in length. The problem is partially resolved by using solid cylinders of constant diameter but different lengths. Although it is true that the shorter ones are stubbier than the longer ones, they are all essentially the same shape, being cylinders of varying lengths but having constant frontal areas.

You are provided with two lengths of dowelling of the same diameter (12 mm, 1/2 inch), two wooden boards, some plasticine, a knife, and a metal rod. Using one of the boards to work on, roll some of the plasticine into a "sausage" that is somewhat thicker than the dowelling. Line up the dowelling on either side of the plasticine, and using the second board on top, roll the plasticine until it is the same uniform diameter as the dowelling. Cut the plasticine rod into lengths of 2, 3, 4, 5, 6, 7, 8, 9, and 10 cm, taking care not to compress the ends. To avoid compression, instead of cutting right through with one slice, cut around the circumference, a little at a time. Round off the ends, and mark the midpoints of each model.

Spike the shortest model onto the metal rod and attach it to the balance, as in the previous wind tunnel experiments. Making sure that the model is parallel to the airflow and that the pointer is set to zero in the usual way, set the airspeed to maximum (14.4 m/s) and measure the drag force. Repeat for the other models and tabulate your results.

How is drag related to length? Are you surprised by the results? Explain them in terms of the way that pressure drag and friction drag act on a body. Remember that we are operating at a fairly high Reynolds number – about 32,000 for a model that is 3.5 cm long. Under these conditions friction drag is negligible and essentially all of the drag is attributed to pressure drag.

Sample Observations (airspeed 14.4 m/s):

Length of rod (cm)	Drag (g)
2.0	5.5
3.0	4.0
4.0	4.0
5.0	4.5
6.0	4.5
7.0	4.0
8.0	5.0
9.0	6.0
10.0	5.0

The drag is essentially independent of length. This is because most of the resistance is attributable to pressure drag, which varies with the frontal area. Friction drag, which acts on the surface of the body parallel to the fluid flow, increases with length. However, since friction drag is so small compared with pressure drag, the effect of increasing the surface area is negligible, at least in this experiment.

The minor role of friction drag, compared with pressure drag, at high Reynolds numbers explains why commercial airliners are essentially cylindrical, and how stretched versions can operate with little increase in drag. It also explains how oil tankers can be enlarged by cutting them in two and welding an additional section in place, like adding a leaf to a table, without necessitating any changes in the power plant.

PART 2   VISCOSITY AND FLUID FLOW, AT DIFFERENT REYNOLDS NUMBERS

The liquid we have most experience with is water, which has a fairly low viscosity. Liquids with high viscosities, like syrup and glycerine, behave quite differently from water under similar conditions. The first experiment will introduce you to some aspects of fluid flow in a high-viscosity fluid. The other experiments are designed to compare the drag forces acting on spheres and on streamlined bodies that move at different Reynolds numbers. The high-viscosity liquid used is glycerine, and although this is perfectly harmless (it is sometimes used as a cooking ingredient), it is horribly sticky. You must take care not to spill it. If you should get glycerine on yourself, or worse, on your clothes, it cannot be wiped off. The only way to get rid of it is with plenty of water. Glycerine is also hygroscopic, meaning that it readily absorbs moisture from the air, so it should be covered up at all times when not in use.

Experiment 1   Investigating the properties of fluid flow in a high-viscosity fluid.

You are provided with glycerine, a glass funnel, a petri dish, a syringe, and some colored glycerine (most colored substances, including potassium permanganate and biological stains, are readily taken up by glycerine).

Fill the petri dish about two-thirds full with glycerine. Invert the funnel into the glycerine (Fig. 11.8). Draw up a small quantity of the colored glycerine into the syringe. Notice how difficult it is to suck up, due to its high viscosity. Fortunately you need only a few milliliters. Carefully draw out a straight line of colored glycerine from the edge of the petri dish to the funnel. Make sure that you continue far enough so that the colored thread touches the side of the funnel. You will find that the thread of colored glycerine remains discrete from the rest.

With equal care, slowly turn the funnel and notice what happens to the colored line. Try reversing the direction of rotation. What happens? Wash out the apparatus and fill the petri dish with water. Repeat the experiment.

The glycerine is so viscous that movement of the funnel causes shearing movements in adjacent layers of the liquid filling the petri dish. As a result, the thread of colored glycerine is drawn along by the rotating funnel and becomes a spiral. When the direction of rotation is reversed, the colored thread tends to

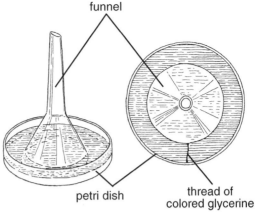

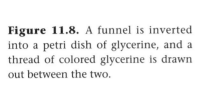

Figure 11.8. A funnel is inverted into a petri dish of glycerine, and a thread of colored glycerine is drawn out between the two.

return to its former position. The rotation of the funnel causes shearing movements at some distance, demonstrating the thickness of the boundary layer. When the experiment is repeated with water, the viscosity is so much lower that there is no shearing of adjacent layers, and the thread of colored glycerine is not spiraled around by the funnel. The only movement of the colored thread is right next to the funnel, showing that the boundary layer is very thin.

Experiment 2   Comparing the drag forces on a small sphere and a streamlined shape when they are sinking in a short column of water.

You are provided with some plasticine, a 250 ml measuring cylinder fitted with a modified stirring rod for recovering models from the bottom, two paper clips, a 250 ml beaker, some lead shot, some thread, and a short length of stiff wire (about 10 cm long).

Break off two pieces of plasticine, each of about 1 cm in diameter, which weigh about 2 g each. Mold one into a streamlined shape, the other into a sphere, removing plasticine from the sphere until both models have the same diameter. The streamlined shape obviously has a greater mass and volume than the sphere.

When a body is immersed in a fluid, it experiences an upthrust equivalent to the mass of fluid displaced (Archimedes' principle). Thus an object weighs less in water than it does in air. For the experiments that follow, it is necessary for the two models to have the same weight when immersed in water. This is achieved by suspending the two models in water and adding lead shot to the sphere until it weighs the same as the streamlined model. The exact procedure is as follows. First make up a simple balance (Fig. 11.9) using the short piece of wire, two paper clips appropriately bent so as to grip the models lightly, and three pieces of thread. Make sure that the central suspension thread is placed exactly in the middle of the piece of wire (which must be straight), and that the distance from the threads supporting the models to the end of the wire is the same for both. Then lower the models into a beaker of water until they are both submerged. Add pellets of lead shot to the sphere until the balance becomes horizontal – that is, until the two

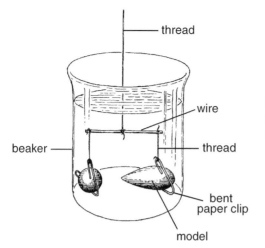

Figure 11.9. A simple apparatus for weighing models in the liquid appropriate to the experiment.

models have the same weight in water. When you have added the correct amount of lead ballast to the sphere, poke the lead shot beneath the surface and smooth off the plasticine.

You now have two models that have the same weight in water. This means that the acceleration force due to gravity will be the same on both models when they are released in water. Any differences in their performance is therefore due to differences in their shape.

Fill the measuring cylinder to the top with water. Holding the two models side by side in the water, release them and see which model reaches the bottom first. You can recover the models from the bottom by using the modified stirring rod. Repeat the experiment until your results are consistent, that is, until the same model reaches the bottom first each time. Since the gravitational force acting on the two models is the same, the difference in their performance is due to the difference in the drag forces acting upon them.

Which model experiences the greatest drag force? Are the results what you would have expected? Probably not. How do you rationalize your results? (Clue: the density of the models is not very high compared with water, therefore they do not sink very fast.)

Experiment 3   Comparing the drag forces on a medium-sized sphere and a streamlined shape when they are sinking in a fairly long column of water.

You are provided with the same apparatus as before, except the measuring cylinder is now much taller (2 liter capacity), and you also have two steel ball bearings (diameter about 1 cm; 1/2 inch).

Using the steel balls and the minimum amount of plasticine, make a second pair of models, one spherical, one streamlined. As before, make sure that the models have the same diameter and the same weight in water.

Fill the measuring cylinder with water and test to see which model consistently reaches the bottom first. A large cylinder is necessary for two reasons: (1) to prevent interference between the two large models; and (2) to give a longer column of water so that the two models may approach their terminal velocity. Ideally, you should use a column two or three times this length to get a good separation between the two falling models.

Your results should be contrary to those of Experiment 2. Why? If your results are not contrary, you have not done a good enough job in making a streamlined shape. Make sure of the following: (1) there is a gentle taper on either side of the shoulder; (2) the leading surface is bluntly pointed; (3) there is a long, gently tapering trailing surface; and (4) the shoulder is about one-third of the way back from the anterior end. Keep modifying your model until the results for this experiment are the reverse of those for Experiment 2.

Without attempting to calculate the Reynolds number in this experiment, comment on how it compares with the Reynolds number in Experiment 2. You should now be able to compare, and discuss, the results of Experiments 2 and 3 in terms of the relative magnitudes of the Reynolds numbers in the two experiments.

Experiment 4 Comparing the drag forces on a medium-sized sphere and a streamlined shape when they are sinking in a fairly long column of glycerine.

Repeat Experiment 3, this time using glycerine instead of water. Make sure that the two models have the same weight, *in glycerine,* at the start of the experiment.

You should find that your results are similar to those of Experiment 2. Rationalize the results in terms of Reynolds numbers and the streamlined body.

Observation 1 Examine the aircraft model supplied, a World War II Spitfire. Notice the streamlined shape of the fuselage.

List any other manufactured structures you have seen, or have knowledge of, where the streamlined shape is used.

Observation 2 Compare the seahorse with the stickleback. They belong to the same taxonomic order, but they differ significantly in form and habits. Notice that the stickleback, like almost all other fishes, has a streamlined body, whereas the seahorse is not streamlined. Based on what you have learned in the experiments, what predictions would you make about their respective swimming performances? (Remember, they both swim in the same fluid – water.)

PART 3 THE BEHAVIOR OF INCLINED PLATES

Experiment 1 Testing the effects of changing the angle of attack.

You are provided with some sheets of paper. Make a paper airplane that has a pair of adjustable inclined plates (control surfaces) at the rear end. You are free to put all of your inventive skills into your design, but if aircraft building is not your

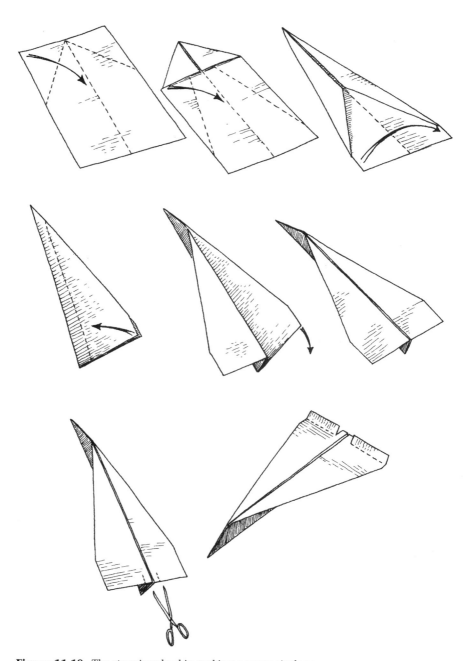

Figure 11.10. The steps involved in making a paper airplane.

forte, you might like to follow the construction plans outlined in Figure 11.10. Determine the approximate position of the center of mass of your aircraft and mark it with a cross. Bend both pairs of inclined plates downward such that the trailing edge is lower than the leading edge – the angle of attack should be about

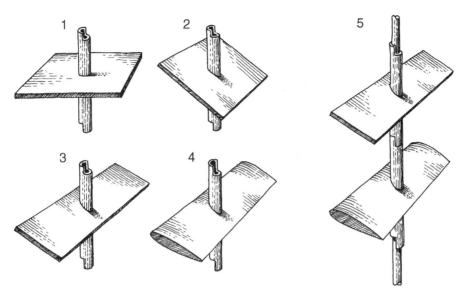

Figure 11.11. A set of inclined plates mounted on Plexiglas tubes that fit onto a Plexiglas rod: (1) square, with zero angle of attack; (2) square, with fixed angle of attack; (3) rectangular plate, with fixed angle of attack and rectangular cross section; (4) the same as (3) but with a streamlined cross section; (5) two inclined plates interdigitated.

20°. When the aircraft is flown, will the inclined plates give an upthrust or a downthrust? What will be the effect of this, acting as it does behind the center of mass? When you have predicted whether the nose will be raised or depressed by the action of the inclined plates, launch your aircraft and test your predictions. I hope you were correct.

Set the inclined plates in the opposite direction and test fly again.

Repeat, this time with one plate set up, the other down. Before launching, predict precisely how the aircraft should perform.

Experiment 2 Testing the effects of changing the aspect ratio.

You are provided with a set of inclined plates and an aquarium filled with water. Each inclined plate is attached to a standard section of pipe with a cutout, which interdigitates with a similar cutout at the bottom of a Plexiglas rod (Fig. 11.11). This mechanism prevents the plate from spinning when it is being pushed through the water. The plates have the same surface area (within tolerable limits) and same angle of attack (except for one of the plates, which has a zero angle of attack). They are as follows: (1) a square plate, with aspect ratio 1, set at a zero angle of attack; (2) a similar square plate, set at a fixed angle of attack; (3) a rectangular plate of similar area and angle of attack as #2, but with an aspect ratio of 4; and (4) the same plate as #3 but with tapered edges.

Thread plate #1 onto the rod and move it through the water at a zero angle of attack. The plate will not experience any lift (make sure the angle really is zero),

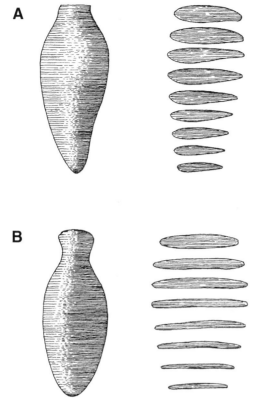

Figure 11.12. A dolphin fin (A), and the cast of an ichthyosaur fin (B), sectioned serially. The dolphin fin, which comprises the soft tissue as well as the skeletal support, has a streamlined profile. The ichthyosaur fin, comprising only the skeleton, has a less obvious streamlined profile.

only drag, which is shown by the force that tends to bend the rod. Now incline the plate at an acute (less than 90°) angle of attack.

When you come to the end of the tank quickly reverse the direction of motion so that the angle of attack is now obtuse (greater than 90°). Notice that the plate is now thrust forcibly downward because an obtuse angle of attack generates a downthrust.

Replace plate #1 with plate #2 (square, fixed angle of attack, aspect ratio 1). Move the plate through the water and observe the upthrust and drag forces when moving with an acute angle of attack.

Add plate #3 (rectangular, same angle of attack, aspect ratio 4). Line up plates #2 and #3 so that they will both move through the water at an acute angle of attack. If the lift/drag ratio of the high-aspect-ratio plate is higher than that of the square plate beneath it, the two plates will move apart when they are moved through the water. In lowering the plates into the water, make sure that they do not float up – use your fingers and twiddle the rod until the two plates interdigitate with one another and with the bottom stop.

Are the results as you had expected? They should be.

Remove plate #2, leaving plate #3 on the rod. Now add plate #4, which is the same as #3 in all respects except that its cross section has been tapered. Do you think that plate #4 has a higher lift/drag ratio than plate #3? Why? Move the plates through the water as in the previous experiment and see whether your predictions are supported.

Observation 1   Examine an electric fan provided. This is an example of an inclined plate. In this application, the desired end result is not lift but airflow. Which way does the fan need to turn to direct air at your face? Test your prediction by plugging in the fan.

Observation 2   Examine the dolphin fin, which has been sectioned transversely (Fig. 11.12A). In life, this fin functions as an inclined plate. Notice that the plate has a moderate aspect ratio. Also notice the streamlined shape of the cross section. Compare this specimen with the sectioned cast of a forefin of an ichthyosaur (Fig. 11.12B). The animal that owned the original structure lived in the sea some 180 million years ago. It faced the same problems of moving through water that the dolphin faces today, and it resolved them in a similar fashion.

Flight and Flying

Our aspirations to fly were inspired by birds, and our earliest attempts to become airborne involved some flapping imitation of a bird's wings. You may, for example, have seen that jerky movie footage of the Frenchman who aspired to fly from the top of the Eiffel tower at the turn of the century. After a few moments of hesitation, he launched himself into the air, furiously flapping his membranous wings like something from a Dracula movie. He failed, of course, not only to understand the principle of flight but also to walk away from his heavy landing. The point missed by this unfortunate soul, and many other aspiring aviators, is that the wing of a bird is more than just a flap to be fluttered. A bird's wing, like that of a Cessna or a bee, has an airfoil – a particular cross-sectional shape – and airfoils have certain characteristics that enable them to generate lift when moved through the air.

Although there is a world of difference in appearance between the wings of birds, insects, and airliners, they almost all work on the same airfoil principle. The fundamental difference among these various fliers is the source of power used to drive the airfoil through the air. Most aircraft have fixed wings that are driven forward through the air by engine power. Most flying animals, in contrast, have movable wings that are driven through the air by muscle power. However, there are some aircraft, called gliders or sailplanes, and some birds, called soarers, that fly on fixed wings, using moving air currents as their source of power.

The Airfoil

Airfoils have a characteristic cross-sectional shape in which the top surface is convex. The bottom surface can be flat (as in many light airplanes, including biplanes), concave (as in most flying animals and our earliest flying machines), or convex (as in some modern airplanes). An exception to the typical airfoil shape is found among supersonic aircraft, whose wings have an angular rather than a curved profile. Since no animal fliers even approach the speed of sound, these specialized airfoils are of no interest here.

If you looked at the wing of most insects you would see that it was perfectly flat. However, when in flight, the pressure of the air on the underside, acting in concert with the asymmetrical arrangement of the stiffening veins,

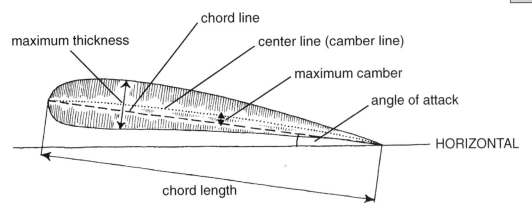

Figure 12.1. An airfoil showing the features discussed in the text.

causes the wing to assume a curved profile (Ennos 1988). Birds, in contrast, have wings with permanently shaped airfoils, often with a thick leading edge and a gently tapered trailing edge, built largely from feathers.

Inclined plates, as we saw in the last chapter, generate lift when set at an acute angle of attack by deflecting fluid from their lower surface. Airfoils, in contrast, generate lift when set at a zero angle of attack by virtue of the convex curvature of their upper surface, which differs in shape from that of the lower surface. Typically, though, airfoils operate at small angles of attack. The reason airfoils generate lift by virtue of their curved profile is because when a fluid flows over a convex surface it experiences a decrease in pressure. You can easily demonstrate this phenomenon by holding a strip of paper to your lips so that the free end droops gently away from you. If you now blow air over the convexly curved upper surface, there is a decrease in pressure. Since the pressure on the lower surface remains unchanged, the paper strip airfoil experiences an upthrust, or lift, and rises in the air.

Hundreds of different airfoil shapes have been devised over the years, each one designed to serve a particular set of requirements. Some, like those of heavy transport aircraft, are thick and heavily cambered for optimum lifting ability. Others, like those of subsonic fighters, are slender and barely cambered, minimizing drag to optimize speed. Several systems have been devised to describe the cross-sectional shape of the airfoil, and one of the most widely used is a four-digit numbering system devised by the National Advisory Committee for Aeronautics in America, referred to as NACA. Before describing how this system works, I need to define certain anatomical features of an airfoil (Fig. 12.1). The *chord line* is a line connecting the center of the leading edge and the center of the trailing edge. The *chord length*, or simply the *chord*, is the length of the chord line. The *center line*, also called the *camber line*, is a line that is equidistant from the upper and lower sides of the airfoil. The *angle of attack* is the angle between the chord line and the direction of airflow (horizontal if the flight path is horizontal). The *maximum camber* is the maximum

Figure 12.2. A specific airfoil, NACA 4412, as used on the Jet Provost.

distance between the chord line and the center line. The *maximum thickness* of the airfoil is the maximum distance between the upper and lower sides. Three variables are used to describe an airfoil section in the NACA system, in the following order:

1. The maximum camber, expressed as a percentage of the chord.
2. The position of the maximum camber, expressed in tenths of the chord, from the leading edge.
3. The maximum thickness, expressed as a percentage of the chord.

For example, NACA 4412, as used on the Jet Provost, a training jet used by the Royal Air Force, is a fairly thick airfoil in which the maximum camber is 4 percent of the chord, it occurs 4/10 or 40 percent of the chord length back from the leading edge, and the maximum thickness is 12 percent of the chord (Fig. 12.2). There are several other numbering systems for describing airfoils; some of these include more information than the four-figure system just described.

Forces Generated by an Airfoil

When air flows over an airfoil set at a zero angle of attack, there is a decrease in pressure on the upper surface and an increase in pressure on the lower surface (Fig. 12.3A). The pressure differential is not the same along the chord, as the peak is displaced toward the leading edge; this is true for both the upper and lower surfaces. Furthermore, the pressure differential is greater for the upper surface than for the lower one; consequently, more lift is obtained for the reduced pressure zone above the airfoil than for the increased pressure below it. You can demonstrate this phenomenon by drilling holes at intervals along the upper and lower surfaces of an airfoil and connecting them to manometers to measure the pressure differentials with respect to atmospheric pressure. Such model airfoils are available commercially and will be used in the practical. These models are also quite easy to make: I made one simply by grinding a block of Plexiglas to an airfoil shape, drilling the holes, and then connecting the airways using a drill press.

The combination of the reduced pressure on the upper surface and the increased pressure on the lower surface results in a net lift, which acts at right angles to the direction of the airflow. This lift force passes through a point on the chord line called the *center of pressure* (Fig. 12.3B). The lift force is always accompanied by a drag force that, acts parallel to the direction of flow and therefore directly opposes forward motion. This drag force actually has two

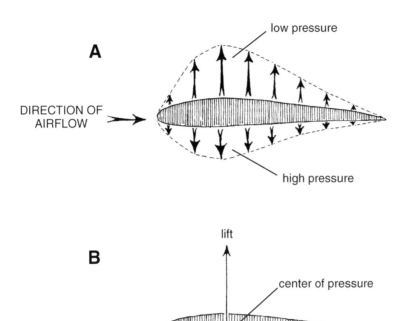

low pressure

A

DIRECTION OF
AIRFLOW

high pressure

B

lift

center of pressure

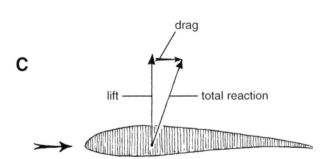

drag

C

lift ——— total reaction

Figure 12.3. The forces generated by an airfoil. When air flows over an airfoil set at a zero angle of attack, there is a zone of low pressure over the top surface and a zone of high pressure over the lower surface (A). The resultant is a net lift, acting at right angles to the direction of the airflow through the center of pressure (B). Lift is always accompanied by drag, acting parallel to the direction of airflow (C). The resultant of lift and drag is the total reaction, which also passes through the center of pressure.

components, referred to as *profile drag* and *induced drag*. Profile drag is caused by the friction drag (or skin drag) of the wing, by its camber, and by any separation of air from its surfaces. Induced drag is caused by wing-tip vortices, which are discussed later (p. 228). The resultant of lift and drag is a vector, referred to as the *total reaction,* which also passes through the center of pressure (Fig. 12.3C). The drag generated by the remainder of the aircraft, or animal, is referred to as *parasite drag* because it is not accompanied by any lift.

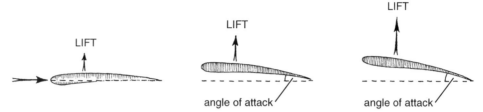

Figure 12.4. The lift generated by an airfoil increases with the angle of attack, up to a point.

The pressure differential that produces the lift force arises from the fact that the air is speeded up over the upper surface of the airfoil and slowed down over the lower surface. These differences in velocities generate a complex system of vortices, collectively referred to as *circulation*. This system is described in the theory of the same name, pioneered by the Englishman F. W. Lanchester (1878–1946). These vortices generate a *downwash,* which may be thought of as producing the lift. Alternatively, the lift may be thought of as being caused by the vortices, with the downwash being a consequence of the lift. The vortices are produced when the airfoil is set at a zero angle of attack, although both forces are increased by having a small angle of attack. For further reading see L. J. Clancy (1975), A. C. Kermode (1972), U. M. Norberg (1990), and D. Stinton (1966).

Whereas the lift force generated by an airfoil increases with increasing angle of attack (up to a point), so, too, does the drag force (Fig. 12.4). The lift-to-drag ratio increases very rapidly up to angles of attack of about 3°–4°, reaching values up to about 24, but then it rapidly falls off (Kermode 1972). As the angle of attack increases still further, a point is reached when there is a sudden decrease in lift, at which point the airfoil is said to have *stalled.* (Stalling can also be brought about by decreasing the speed of the air passing over the airfoil.)

At this point I would like to distinguish between *airspeed* and *ground speed,* concepts that are important to aircraft, both animate and inanimate. Airspeed is the speed of the air relative to the aircraft. Ground speed is the speed of the aircraft relative to the ground. An aircraft flying at a ground speed of 450 mph with a tail wind of 40 mph would have an airspeed of 410 mph. Conversely, a bird flying into a strong head wind of 30 mph could have an airspeed of 20 mph – enough to keep it flying – but a negative ground speed of 10 mph. As far as the bird is concerned it is flying forward relative to the air that surrounds it, but relative to the ground below, it is flying backward.

The airspeed at which stalling occurs is referred to as the *stalling speed.* Stalling is caused by the separation of the airflow from the upper surface of the airfoil. Separation begins at the trailing edge and moves forward, and as it does so, the lift decreases. Most of the lift is lost at the stalling point, and when lift is no longer large enough to support the weight of the aircraft, it loses height. Pilots and flying animals alike therefore avoid allowing themselves to get into a stalling situation during the normal course of flying. The only time when stalling is desirable is during landing, which is essentially a controlled stall. However, most modern aircraft are flown onto the runway with the wings in an unstalled condition. Lift is then quickly lost by the deployment of air-spoilers and reverse engine thrust. You can check this for yourself if you fly on an aircraft like the Canadian de Havilland Dash series, in which the cabin is slung beneath the wings. Choose a seat where you can watch the undercarriage. On landing you will see that the springing system of the undercarriage only becomes compressed by the weight of the aircraft some little time after the wheels have touched the runway. The reason for touching down at speeds well above the stalling speed is to allow for a margin of safety in the event of wind gusts or the need to abort the landing at the last moment and take off again. The landing of birds, in contrast, is a controlled stall, and sometimes, if you are very lucky, you can catch a glimpse of the wing of a bird actually stalling as it comes in to land. Pigeons are the most likely candidates, not only because they are common but because they have a low landing speed. The place to watch is the top surface of the wing. At the point of stalling, any loose feathers that were being held in place by the airflow suddenly pop up as the airflow separates from the surface. In several years of pigeon-watching I have seen the feathers pop only once or twice, but I am told that you can always see this when crows are landing. This same method of capturing the instant of the separation of the airflow is often used during the testing of aircraft. Tufts of wool are taped over the surfaces of interest and the aircraft, or a model, is then tested in a wind tunnel or flown and monitored by observers in a chase plane.

High-Lift and Antistalling Devices

The lift force of a wing at a given angle of attack can be increased by increasing its camber, its thickness, and its surface area. However, improvements in lift are always accompanied by increases in drag, which means more energy is required to push the airfoil through the air. A compromise strategy employed by modern airliners is to have a variety of high-lift devices to increase the camber and surface area of the wing temporarily, prior to takeoff and landing. These devices take the form of various leading and trailing edge flaps that are retracted when the aircraft is in level flight.

Some aircraft also employ an antistalling device. The first such device to be patented was the Handley Page leading edge slot (for a good historical account see Lachmann 1964). Conceptually, this was a slot cut spanwise along the leading edge of the wing, designed to duct high-pressure air from the lower surface into the zone of separation on the upper surface (Fig. 12.5). In its simplest practical form, as used in aircraft built before World War II, the leading edge slot was just a narrow slat, attached to the leading edge of the wing and

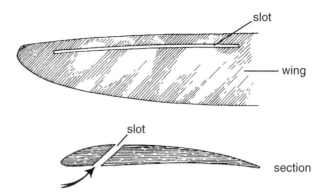

Figure 12.5. A leading edge slot ducts high-pressure air from the lower surface of the wing to the upper surface, as shown in the section through the wing.

separated by a distance of a few centimeters (the gap was the slot). By redirecting the flow of air over the top surface of the wing, the leading edge slot delayed the onset of separation. In later aircraft, including the Messerschmitt 109, the leading edge slat was flush-fitting but popped out automatically just before the wing stalled, thus providing a temporary slot in the leading edge of the wing. An alternative form was to have a slot, or slots, cut into the leading edge of the wing, as in the early conceptualization. This type, referred to as a slotted wing, was used on the rocket-propelled Messerschmitt 163. Incidentally, one of the former technicians in our department, onetime Luftwaffe pilot Rudy Zimmermann, used to fly this remarkable little aircraft. He told me that it was impossible to stall the aircraft, no matter what the pilot did with it.

Birds have a similar device, called the *alula,* or bastard wing, which is the bird's thumb. The alula has the form of a small wing on the leading edge of the wing, about halfway along its length, where the wing bends. If you look carefully at a pigeon when it is coming in to land, you will see the alula pop out, just before the bird touches down. There was a time when I would see an alula deployed only once or twice in watching dozens of pigeon landings. However, my eyes are now well practiced in where to look, and it is a rare morning when I do not see at least one alula deployment. The alula is relatively small compared with the length of the wing, but it appears to be located in the region of the wing that is most likely to stall. It therefore serves a useful role in spite of its small size. By delaying the stalling of the wing, the alula reduces the stalling speed, and hence the landing speed, of the bird, thereby making for a more carefully controlled descent.

Flaps, Slots, and Slats

The various high-lift devices on the leading and trailing edges of the wing of a modern airliner are often referred to generically as *flaps.* Flaps are usually, but not always, attached to the trailing edge. For example, the Boeing 747 has a

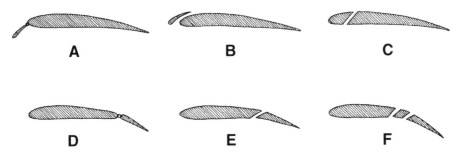

Figure 12.6. A selection of some of the different types of flaps and slots, as used on the wings of various aircraft: (A) Krueger flap; (B) fixed slat; (C) slotted wing; (D) plain flap; (E) slotted flap; (F) double-slotted flap.

hingelike flap on its leading edge that swings down like a narrow door (Fig. 12.6A). This type of flap is referred to as a Krueger flap. If the device on the leading edge is separated from the leading edge of the wing by a gap, it is referred to as a *slat,* and the gap is referred to as a *slot* (Fig. 12.6B). Permanent spanwise slits along the leading edge of the wing, as in the Messerschmitt 163, are also referred to as slots, and the wing is referred to as a *slotted wing* (Fig. 12.6C).

Most airliners have an array of flaps that are extended from the trailing edge prior to takeoff and landing (check for yourself next time you fly). These may simply be hinged, which is referred to as a *plain flap* (Fig. 12.6D). Often the flap, once fully extended, is separated from the wing by a gap (slot), and this type is referred to as a *slotted flap* (Fig. 12.6E). If the flap has a double slot it is called a *double-slotted flap* (Fig. 12.6F).

Wing Characteristics

The lift-to-drag ratio is a useful measure of the efficiency of a wing. Wings with high aspect ratios have a higher lift-to-drag ratio than do short, wide wings. One of the consequences of this fact is that the *gliding angle,* that is, the minimum angle at which the bird can fly during gliding, is shallower for those birds having wings with a high aspect ratio. Gull wings have higher aspect ratios than pigeon wings, which explains why the gulls can glide at much shallower angles. Having a low gliding angle means that a bird can travel longer distances for a given loss of height. Therefore, if a gull and a pigeon both glided down to street level from the same height, the gull would travel further.

Weight also has a considerable influence on flying performance, and it is expressed in terms of *wing loading,* that is, the total weight of the aircraft divided by the surface area of the wings. Aircraft with low wing loadings have lower stalling speeds and are therefore more maneuverable because they can make tighter turns, just as a car can when it is traveling more slowly. They also consume less energy and have lower sink rates (rate of vertical descent) and shallower gliding angles than aircraft with higher wing loadings.

A useful way of seeing the connections between aspect ratio, wing loading, power, and flying performance is to compare two famous fighters of World War II: the Messerschmitt 109 and the Supermarine Spitfire. There were many different versions of each aircraft during the war, and the data used here were based upon the first operational models: the Messerschmitt 109 E-1, and the Spitfire I. Apart from the Messerschmitt's slightly shorter wingspan (9.8 m or 32 ft compared with 11.3 m or 37 ft) and marginally greater weight (2,455 kg or 5,400 lbs, compared with 2,400 kg or 5,280 lbs), the two aircraft were very similar in size. The Messerschmitt, however, did have a somewhat larger engine; its Daimler-Benz delivered 1,100 horsepower compared with the more modest 880 horsepower of the Spitfire's Rolls-Royce power plant. The most striking difference between them was the shape of the wings. The Spitfire had much broader, elliptical wings, giving a lower aspect ratio and a wing area almost 40 percent greater than that of the Messerschmitt. As a consequence, the wing loading of the Messerschmitt was 33 percent higher than the Spitfire's (154 kg/m^2 or 32 pounds per square foot, compared with 115 kg/m^2 or 24 pounds per square foot). This conferred a higher stalling speed on the Messerschmitt. It could climb and dive faster than the Spitfire, and it had a higher operational ceiling. However, because of its higher stalling speed, it could not fly as slowly and thus could not make such tight turns, making it less maneuverable.

Wing-Tip Vortices and Aspect Ratios

Given that an airfoil generates lift by differential air pressures, it follows that air flowing over the top surface of a wing is below atmospheric pressure, whereas that flowing over the lower surface is above it. Visualize looking down on the top surface of the wing of an aircraft in flight. Because the air that surrounds the upper surface of the wing is at a higher pressure, it tends to encroach upon the wing, causing the airflow over the wing to be deflected inward, away from the wing tips (Fig. 12.7A). Now consider the underside view of the wing. This surface of the wing is surrounded by lower pressure air, so that the airflow over the wing's lower surface tends to be deflected outward, toward the wing tips (Fig. 12.7B). Since the air beneath the wings is at a higher pressure than that above it, there is a continuous flow of air around the wing tip, from the bottom surface to the top. As the flow is swept aft by the aircraft's forward motion it is given a spiraling motion due to the opposing directions of flow over the top and bottom surfaces, resulting in wing-tip vortices. The wing-tip vortices are the cause of induced drag.

High-aspect-ratio wings have a higher lift-to-drag ratio than low-aspect-ratio wings. One explanation for this is that a high-aspect-ratio wing has a smaller wing-tip area relative to the area of the rest of the wing. Consequently, the wing tips spill relatively less air from the lower surface of the wing tip, thereby losing less lift and generating smaller wing-tip vortices (Kermode

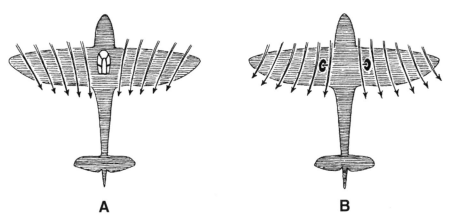

A   **B**

Figure 12.7. The airflow over the top (A) and bottom (B) surfaces of the wings of an aircraft is influenced by the pressure differential of the surrounding air.

1972). An alternative explanation is that higher aspect-ratio wings have wing-tip vortices that are farther away from the most efficient part of the wing (the middle). Hence the unfavorable influences of the wing-tip vortices are reduced.

Yaw, Pitch, and Roll

In order to fly or swim successfully, a body must be able to compensate for alterations in its path caused by fluid movements, and it must be able to change course at will. There are three basic movements that can occur: yaw, pitch, and roll, and I will illustrate these by reference to a simple aircraft (Fig. 12.8). *Yaw* is a side-to-side turning movement about a vertical axis, *pitch* is an up-and-down rotational movement about a horizontal axis, and *roll* is a turning movement about a longitudinal axis. All three movements are corrected for during the flight of an aircraft by the action of control surfaces. Yaw is corrected by the action of

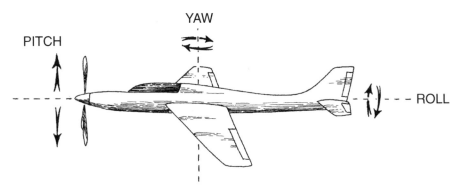

Figure 12.8. Movements in the three planes of space.

the vertical rudder, or fin. When an airplane is flying on a straight course the rudder cuts through the air in the direction of travel, but when the machine begins to yaw, the rudder becomes an inclined plate set at a small angle of attack, and the deflection force generated brings the airplane back onto a straight course. Control of pitch is brought about by the action of the horizontal stabilizer, which similarly acts as an inclined plate. The flight feathers of a dart or an arrow correct for yaw and pitch in the same manner; if a dart is thrown with a wobble, it corrects itself very rapidly and flies straight. In addition to correcting for yaw and pitch, the tail of an airplane has adjustable control surfaces that can *cause* it to yaw (turn) and pitch (climb or dive). Roll is effected by the ailerons, which are elongated flaps on the trailing edges of the wings, placed toward the wing tips. Increasing the angle of attack of an aileron gives an additional lift to the wing on that side, making the aircraft roll toward the other side.

Soaring and Gliding

Large birds generally have difficulty becoming airborne; their takeoffs involve a long, flapping run. Their exertions require the additional power of anaerobic muscle fibers, but they cannot keep up anaerobic metabolism (sprinting) for long because it incurs an oxygen debt. Consequently, large birds are not flapping fliers; they resort to flapping only during takeoff and landing and for brief bursts during normal flying. Instead of using muscle energy for flying, they extract energy from the environment, specifically from moving air masses. This strategy is called *soaring*. The term *gliding* is often used interchangeably with soaring, but the two terms are not synonymous. Gliding is a passive mechanism in which height is continuously lost, whereas in soaring, height is gained or maintained. Soaring flight is considerably less energetic than flapping flight, and the savings increase with body size. It has been estimated that a large bird like a stork consumes about twenty-three times more energy when flapping than when soaring, compared with only about twice as much energy for a small bird like a warbler (Pennycuick 1972). Soaring is not without costs, though. Wind tunnel experiments on gulls have shown that their metabolic costs during soaring rise by a factor of about two over their resting levels (Baudinette and Schmidt-Nielsen 1974). This increase, presumably, is attributable to the muscle activity involved in adjusting the position of the wings, tail, and other parts of the body during soaring flight.

Flapping Flight

Although soaring birds extract energy from the environment to keep them aloft, they all use flapping flight to become airborne. They also occasionally use flapping flight once they are in the air, to change directions, to fly toward new thermals, and for other short-term purposes. Flapping flight is therefore universally present among birds, and it is the sole flying mechanism for most of the

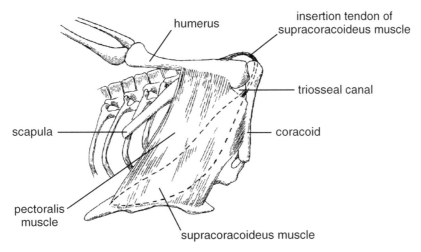

Figure 12.9. The supracoracoideus muscle (dotted) lies deep to the pectoralis and brings about an elevation of the humerus because of the way its tendon of insertion loops through the triosseal canal.

smaller ones. While differing in detail among species, a bird's flapping flight mechanism involves a downward and forward power stroke with the wing fully extended, followed by an upward and backward recovery stroke with the wing partially retracted. As the wing has an airfoil cross section, when the bird drives it through the air during the power stroke, it generates a lift force directed obliquely forward and upward, counteracting both drag and gravity. The upward wing beat is largely a recovery stroke, though most birds probably generate some lift and thrust when flying slowly, such as when they take off. The muscle power for the wing movements is mostly provided by the breast muscles attached to the sternum, a deeply keeled structure in flying birds.

There are two major pairs of wing muscles powering the up-and-down movements of the wings. Both pairs originate from the keeled sternum, with one pair lying deep to the other (Fig. 12.9). The outer muscle, the pectoralis, is much larger than the underlying supracoracoideus muscle. The pectoralis muscle powers the downstroke, whereas the supracoracoideus, which is usually only about one-tenth as large, is generally considered to power the upstroke. That two muscles can effect diametrically opposed wing movements, even though they occupy similar positions, is because of the way their insertions are attached to the humerus. The insertion tendon of the pectoralis major connects directly to the leading edge of the proximal end of the humerus, so when the muscle contracts, it pulls the wing downward and forward. The insertion tendon of the supracoracoideus, in contrast, takes a loop through a bony canal in the shoulder region, the *triosseal canal* (foramen triosseum), and inserts onto the dorsal aspect of the proximal end of the humerus. Consequently, when the smaller muscle contracts, it elevates the wing in preparation for the next downstroke. However, a recent study conducted on the starling (*Sturnus vulgaris*) shows that the primary role of the supracoracoideus may be to rotate the humerus (Poore, Sánchez-Haiman, and Goslow 1997).

Once a bird is flying, its forward motion is added to the forward motion of its wings during the power stroke; this increases the airspeed over the wing. Prior to takeoff, though, the airspeed over its wings is zero (unless the bird is heading into the wind), and lift can be achieved only by raising this airspeed above the stalling speed. Small fliers, like small birds, bats, and insects, have no difficulty generating this lift, as evidenced by how readily houseflies and sparrows take to the air. Most large fliers, though, have to run or jump while flapping their wings to become airborne. These differences in flight performance are directly attributable to size differences.

Flying Ability and Body Size

As objects get larger their volumes and masses increase with the cube of their length, whereas their areas increase only with the square (Chapter 1). Consequently, large animals have higher wing loadings than do small ones (wing loading = mass/area = l^3/l^2, therefore wing loading $\propto l$). For example, the wing loading of a bee is about 1 kg/m^2 (3.3 ounces per square foot), compared with 5.1 kg/m^2 (1.1 pounds per square foot) for a starling and 8.6 kg/m^2 (1.8 pounds per square foot) for a pelican.

Stalling speed is directly proportional to wing loading, specifically to the square root of wing loading, so larger animals have higher stalling speeds and require higher airspeeds for takeoff. Insects and small birds achieve these airspeeds simply by flapping their wings rapidly, which explains how hoverers can fly. A bee, for instance, can flap its wings more than 200 cycles per second, hence the low-pitched buzz, but a starling cannot do much better than about five beats a second. Many large birds take off by running and flapping and taking advantage of prevailing winds, whereas others become airborne by launching themselves from trees and other high places. But being big does not necessarily mean having takeoff problems. Some fairly large birds of prey, like vultures, have wings with high lifting capacities, mainly due to their large size and to their shape.

Small fliers, and this includes most flying insects, small birds, and all but the largest bats, have no difficulties in taking off. Indeed, many flying insects and hummingbirds have so much power available for flight that they can hover at will. At the other end of the spectrum are large birds like albatrosses and condors that have a running, flapping struggle to take off. Once airborne, though, their flight becomes effortless, but this is because they extract energy from the environment instead of using muscle power, as we will see later.

The perception that flight becomes progressively more difficult with increasing body mass has led to speculations that there is some intrinsic size limitation to flight, related to power. Specifically, it has been proposed that the power requirement for sustained flight rises at a rate slightly higher than the increase in body mass, whereas the muscle power available increases according to body mass$^{2/3}$ (see Marden 1987 for a discussion). It follows from this reasoning that there is an upper size limit for sustained flight, and this limit is believed to correspond to a body mass of about 12 kg (26 pounds). Several living birds are of this mass, including the Andean condor (*Vultur gryphus*), the mute swan (*Cygnus olor*), the white pelican (*Pelecanus onocrotalus*), the wild turkey (*Meleagris gallopavo*), and the kori bustard (*Ardeotis kori*). An underlying assumption of this belief is that the flight muscles available to power the wings are a relatively constant percentage of the body mass. To express this relationship in aeronautical terms, the assumption is that the power-to-weight ratio for fliers is essentially constant, meaning that the engine power keeps step with the mass of the aircraft.

How good is this assumption? J. H. Marden (1987) devised an experiment in which fliers, ranging in size from small insects to the pigeon, were progressively loaded with attached weights to see the maximum mass they could lift. He later dissected each animal to determine the mass of its flight muscles and how that related to the lift force generated and to the body mass. He found that the maximum lift force generated per kilogram of flight muscles varied very little among the 190 species assessed, ranging from 54 to 63 N/kg. This is quite remarkable considering that the species ranged from butterflies and bees to bats and robins. However, there was a considerable range of variation in the ratio of flight muscle mass to body mass. The ratio ranged from 0.115 to 0.560 (data from other studies were also included) and was correlated with flying performance. Animals whose flight muscles comprise less than about 16 percent of their total body mass appear unable to take off from a standing start. However, birds at the top end of the range can take off vertically without any trouble. The coot, for example (a member of the rail family, noted for their poor flying ability), has relatively small flight muscles. These comprise only about 13 percent of its body mass, compared with 22 percent for a pigeon. The coot lives beside lakes and ponds and has to take a long flapping flight across the water to become airborne, but the pigeon can take off vertically from a standing start. Condors have a small mass of flight muscles relative to their large body mass and have some difficulty becoming airborne. However, the North American wild turkey, which is as heavy as a condor, can take off almost vertically from a standing start, by virtue of its much larger flight muscles.

What does all this mean? First, it is apparent that flying animals do not all have the same power-to-weight ratio. Being a large flier therefore does not necessarily imply having difficulties becoming airborne, as shown by the wild turkey. The idea that there is an intrinsic size limitation on flight due to power availability therefore seems uncertain. That is not to suggest that there is *no* upper size limit for flying animals, only that the 12 kg limit as exemplified by today's birds is probably an artifact that has nothing to do with some power-related barrier. This is supported by the fossil remains of *Teratornis*, a

Figure 12.10. The wings of a sprinter, like a woodcock (A), have a lower aspect ratio and are more highly cambered than those of an endurance flier, like a duck (B).

Pleistocene bird of prey with an estimated wingspan of 5 m (16 feet). C. P. Ellington (1991) has suggested that a more likely size-limiting factor than muscle power is the lift that can be produced by wing flapping. Since the flapping rate of an animal's wings is inversely proportional to its body length, large birds are unable to flap their wings very fast. Consequently, there is a limit to the velocity of the airflow over the wings that can be achieved by flapping. Therefore, as animals get larger, a point is reached where the wings cannot be flapped rapidly enough to generate sufficient lift for takeoff.

Sprinters and Cruisers

The North American wild turkey belongs to the Order Galliformes, which includes game birds such as grouse, partridge, and ptarmigan. They were called game birds because of the "sport" they provided the hunter; their fast takeoffs make them more challenging targets. Although they can fly very fast for short distances, they lack endurance, like other sprinters. As we have seen (Chapter 9), part of the reason for their rapid acceleration is that their flight muscles are powered by fast-glycolytic fibers (white muscle). We have also seen that birds with a good takeoff performance have a high value for the ratio of flight muscle mass to body mass. A third factor concerns the shape of the wings. Sprinters have short, broad wings that are often highly cambered (Fig. 12.10A). Such low-aspect-ratio structures are adapted for moving large volumes of air, which optimizes acceleration, but the costs in terms of drag and muscle power are very high. Endurance fliers, like ducks, in contrast, have wings with much higher aspect ratios that may be less highly cambered (Fig. 12.10B). The wings are also powered by muscles that predominate in fast-oxidative/glycolytic fibers (red muscle). The maximum power output for aerobic muscles is about 100 watts/kg, and this is true for vertebrates and insects. As we saw earlier (Chapter 9), anaerobic muscles generate more power than aerobic ones, and for birds the factor is about 2 to 2.5. The factor is even higher for reptiles – up to about 4.5 – which has implications for the flight of the giant pterosaur *Quetzalcoatlus,* an animal that had an estimated wingspan of about 11 meters (36 feet). *Quetzalcoatlus* is the largest flying animal known, and if its flight muscles comprised fast-glycolytic fibers, this would maximize the power available for takeoff. The flapping rate of its wings would have been low, but their large size would presumably have generated sufficient thrust by accelerating a large volume of air.

From Small Flappers to Large Soarers: Summing Up

As animals get larger their wing loadings increase and their wing-beat frequencies decrease. This relationship has profound effects on flying performance. At one end of the spectrum are small maneuverable flappers, like hummingbirds, that can hover but not soar. Middle-sized flappers like pigeons take off readily and can soar, briefly, but not hover. Large birds, like pelicans, soar, with only intermittent flapping. They are not so maneuverable and often have difficulties taking off. Takeoff performance is not entirely dictated by size, though, and there is a strong correlation between performance and the percentage of flight muscle relative to body mass. Thus some large birds, like the wild turkey, have no difficulties becoming airborne, by virtue of their large flight muscles.

Although power output from flight muscles may not impose such a low size ceiling as previously supposed, there may be structural limitations on how large fliers can become, partly due to the need to minimize mass. Aeronautical engineers were faced with the same problem, but it was resolved by the invention of new materials and new technologies. The Wright brothers' first biplane, for example, which was made of wood, wire, and canvas, was driven by a piston engine and weighed only 275 kg (605 pounds). The largest aircraft flying today, the Antonov An-225, is constructed of lightweight alloys, plastics, and composites, is driven by gas turbines, and has a staggering maximum takeoff weight of 660 tons (Matthew 1994). Technology has emancipated the airplane from the constraints of size, but living organisms are still made of the same materials and powered by the same muscles that evolved millions of years ago. These constraints ensure that a paleontologist will never dig up the remains of a flying animal as large as a DC9, far less that of a 747.

Wing Structure in Birds

The wings of modern aircraft are impressive structures, with their leading edge slats and slotted trailing edge flaps, but they are crude devices compared with the aerodynamic elegance of a bird's wing. Just to hold a bird's wing – to feel its lightness and stiffness, check its camber, and see how the feathers glide past each other without leaving gaps when the wing extends and retracts – fills

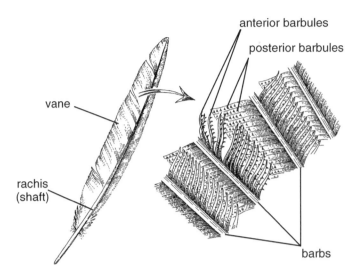

Figure 12.11. The structure of a bird's contour feather.

me with wonder. We could not build a variable geometry wing as light and stiff and as aerodynamically sound, so how is this feat achieved in birds? The secret lies in the structure of the feather.

Try holding a feather in your fingers, but find a regular stiff feather (contour feather), not one of those fluffy down feathers used to stuff pillows (Fig. 12.11). If you hold the feather by the quill, which is the lower end of the *rachis* (shaft), you will notice that it is round and hollow. If you try squeezing it, it will not collapse. While still holding the shaft between your finger and thumb, check its stiffness by applying gentle pressure with another finger. It is remarkably stiff. The shaft, and the rest of the feather, is constructed from the protein keratin – the same material as hair and nail; it is tough and has a low density. Notice that the feather's body, the *vane*, is formed from numerous fore and aft side branches from the shaft, called *barbs*. Adjacent barbs are connected together along their entire length. Run a finger along the vane and you will see that it is quite flexible and will distort. The barbs remain locked together, though, so when you remove your finger the vane springs back to its original shape. The barbs can be separated by pulling them apart, which tears the vane. But you can reverse this by stroking the vane between two fingers. Birds do this for themselves when preening their feathers with their bills. This practice works because of the way the barbs are joined together. Examine the barbs with a lens and note their side branches, called *barbules*. The anterior barbules are armed with hooks, whereas the posterior ones have ridged plates onto which the hooks of the adjacent barbules latch, just like Velcro. The elegance of the design, combining stiffness and strength with lightweight construction, can be appreciated by holding a feather up to the light and seeing how much empty space there is. Indeed, the large volume of air entrapped by feathers accounts for their excellent insulatory properties.

The vane has an airfoil cross section that is convex on the upper surface and concave below. Furthermore, the wing's contour feathers have their shafts displaced toward the leading edge, just as the main spar of an aircraft wing is displaced toward the leading edge. This configuration aligns the supporting structure along the center of lift, so that the lift force does not cause twisting. The feather, then, is a miniwing and can generate lift when air flows over its surface. Having the shaft displaced toward the leading edge also has important functional consequences during the wing-beat cycle. During the downstroke (power stroke), when air is being forced against the underside of the wing, the asymmetry of the position of the shaft causes the feathers to rotate, such that the longer trailing edge is elevated. Since the trailing edge of one feather underlaps the adjacent feather, they are pressed tightly together. This prevents the air from pushing its way between the feathers, which would reduce the efficiency of the power stroke. The reverse occurs during the recovery stroke. The air pushes down against the dorsal surface of the wings, causing the feathers to rotate such that their trailing edges are depressed, opening up gaps between adjacent feathers. This permits air to pass through the wing, thereby reducing the resistance to the upward stroke.

Many birds, especially those that soar, splay their terminal feathers, giving the wing tips a ragged appearance. Crows immediately come to mind as one of the common exemplars of this feature. Quite often these terminal feathers are *emarginated,* meaning that the distal portion is narrower than it is proximally, so the wing tip is permanently splayed, like spreading fingers. This type of wing tip, referred to as *slotted,* reduces induced drag during flight by reducing the extent of the wing-tip vortices, thereby raising the wing's lift-to-drag ratio (Tucker 1993). This device also aids takeoff because when the wings are flapped, the individual feathers function as individual high-aspect-ratio miniwings, contributing to lift. This was clearly demonstrated in a study among galliform birds, where those with more markedly emarginated wing tips were shown to have the highest vertical takeoff performance (Drovetski 1996). Just as birds inspired human flight, so their slotted wing tips attracted the attention of aeronautical engineers. When the oil crisis hit the Western world during the 1970s, experiments were conducted with finlets attached to the wing tips of a Boeing 707, to see if they improved fuel economy. The results were encouraging: fuel consumption dropped and takeoff performance improved. Wing-tip devices are now a standard feature of several modern airliners, including the Boeing 747, the Airbus A-320, and the McDonnell Douglas MD-11. However, it should be pointed out that these devices are placed essentially at right angles to the span of the wing, rather than parallel to the span as are emarginated wing feathers; aerodynamically they function differently.

As we saw in Chapter 6, the wing bones of birds utilize the same light-weight tubular construction as do the feathers. Bats similarly have thin-walled tubular bones in their wings. Experiments on flying bats have shown that torsional stresses in the humerus and radius are considerably higher in fliers than in terrestrial animals (Swartz, Bennett, and Carrier 1992). This is because the wing bones are positioned near the leading edge, whereas the wing's center of lift lies farther back. Consequently, the wings tend to twist during flight. Torsional forces are best resisted by large-diameter tubes – another functional reason for the tubular construction of wing bones.

Although the bony skeleton supports the most anterior part of the wing, much of the leading edge is formed by an elastic web of skin, the *propatagium,* that extends from the shoulder to the wrist. Next time you see poultry being prepared for the oven, take a look at the propatagium and see how large and remarkably resilient it is. The propatagium probably functions as a leading edge flap, and is supplied with muscles that might be able to modify its shape.

Aircraft have devices for measuring airspeed and attitude, giving the pilot warning of the separation of the boundary layer from the wings' upper surface prior to stalling. Birds also appear to have such devices in the form of mechanoreceptors at the bases of their wing feathers (Brown and Fedde 1993). These receptors respond to feather movements by changing their rate of firing of nerve impulses to the brain. The receptors associated with the feathers of the propatagium and alula are believed to give warning of an imminent stall. Those of the secondary feathers (the ones attached to the ulna) might be able to measure airspeed by detecting rate changes in the vibration of the feathers in the airstream.

Soaring Flight

One of the most common sources of moving air exploited by soaring birds are thermals, which are the rising columns of warm air caused by the differential warming rates of different substrates. A large rock in the middle of a savannah, for example, will warm up beneath the sun more rapidly than the surrounding grassland, sending up a vertical column of warm air. Thermals are usually quite narrow, often only a meter or so in diameter, and fliers have to make tight turns to stay within them. Such maneuverability requires slow flying speeds. Soaring birds that exploit thermals therefore require low wing loadings. Most soarers also have low-aspect-ratio wings, which allow for tighter turning circles than high-aspect-ratio ones. Slotted wing tips also improve their flight performance.

Thermals are not restricted to narrow vertical columns because they can be displaced horizontally by winds. This displacement gives rise to soarable corridors, called *thermal streets,* that birds can use for cross-country flights (Pennycuick 1972). Thermals can also take the form of large ascending bubbles, up to about 2 km (1 mile) in diameter. As long as a flier stays within the bubble it can fly in any direction and still gain height. A strategy used by soaring birds, and glider pilots, is to ascend on a thermal to gain height, leave it and glide across country losing height, then ascend on another one. Birds are as adept at recognizing likely places for thermals as experienced glider pilots, and this ability was graphically demonstrated to C. J. Pennycuick

(1972). While studying soaring birds in East Africa with a glider, he noticed that their flight paths took them through more thermals than would be expected if they had been flying randomly. On one particular flight he decided to follow a vulture, and he made much better time by doing so than on his return trip when he had to rely on his own skills in locating thermals.

An entirely different source of moving air exploited by soarers, primarily seabirds, arises when horizontal winds are deflected upward, as when onshore winds blow against sea cliffs. As the air masses are not confined, maneuverability is of less importance. The primary consideration is efficiency – maximizing the lift-to-drag ratio. This is achieved by employing high-aspect-ratio wings, like those of gulls, which are naturally associated with sea cliffs. Their long slender wings have a higher lift-to-drag ratio than the much lower aspect-ratio wings of thermal soarers like vultures. The kind of soaring exemplified by gulls and practiced by many other seabirds is described as *slope soaring*, for obvious reasons. As the zone of rising air follows the coastline, seabirds are able to patrol back and forth along the coast at will. Slope soaring is not confined to cliffs, because when the wind blows against waves, especially when a heavy sea is running, it generates updrafts that seabirds can exploit. A few years ago I was on a Canadian research vessel fishing for swordfish off the coast of Nova Scotia. Gulls were the most common birds at first, but they became scarce when we got farther out to sea. Most of our fishing was done on the edge of the continental shelf, about 160 km (100 miles) from shore. When the wind began to blow and the waves began to grow, along came the petrels, shearwaters, and fulmars – members of the Order Procellariiformes, or tube-nosed birds. Storm petrels, the smallest ones, are sparrow-sized and fly close to the waves with a skittish fluttering of their wings, sometimes hovering, sometimes dabbling with their feet on the water. But the larger procellariiforms are adept slope-soarers, skimming across the tops of the windblown waves with barely a flap of their gracile wings.

Albatrosses are the largest of the procellariiforms. The largest species, the wandering albatross, has a wingspan of 3.4 m (11 feet), which is the largest of all living birds. They commonly slope-soar (Pennycuick 1982) like their smaller relatives, but they apparently also practice dynamic soaring, exploiting the differences in the horizontal wind velocities at different heights above the sea – that is, the boundary layer. They make a shallow dive into the wind from a height of about 20 m (60 ft). As they lose altitude they gain ground speed (speed relative to the ground). At the bottom of the descent, just above sea level, they make a turn into the wind, and the ground speed is so high that they ascend rapidly. As they climb higher the ground speed slows, but because the air encountered is moving progressively faster, the airspeed does not diminish so rapidly. As a consequence, they continue to climb, even though their ground speed is low, until they get high enough to repeat the process. The extensive wanderings of the albatross across the trackless southern oceans is almost legendary. Although it has long been surmised that they make extensive feeding forays during the breeding season, this has now been verified by satellite tracking (Jouventin and Weimerskirch 1990). Each member of a breeding pair takes turns looking after the nest while the other heads off to sea in search of food. During the investigation, conducted on a breeding colony in the Crozet Islands in the Indian Ocean, radio transmitters were attached to five males, and their movements were tracked by satellite. Their feeding forays, lasting two to five weeks, covered distances ranging from 3,664 km (2,257 miles) to a remarkable 15,200 km (9,363 miles). The flying was mostly done in daylight, at speeds up to 80 km/h (49 mph), but the birds were still active at night, especially during moonlight. Most of the time was spent on the wing, though they did land on the ocean's surface for brief rests, which never exceeded an hour and a half.

A few years ago the Royal Ontario Museum obtained an albatross from New Zealand, and I showed it to my class. Its body was pretty impressive – about a meter long and as plump as a goose – but when we spread out the wings the tips almost touched the walls of our small laboratory. We were all suitably impressed, as much by the wings' slenderness as by their length. Later, when the internal organs were removed, I examined the stomach contents and found it crammed full with horny beaks of squid. How many lonely miles of the South Pacific had that magnificent bird patrolled to collect such a bountiful harvest?

High-aspect-ratio wings are energy efficient; they enable an albatross to cover enormous distances at minimal cost. The disadvantage is that they tend to have higher wing loading – this is certainly true for the albatross, which is one of the largest flying birds – and therefore higher stalling speeds. This requires higher takeoff and landing speeds, often resulting in some inelegant mishaps that may have prompted the epithet "gooney bird." As depicted in wildlife movies, albatrosses approach their takeoffs with the apprehensions of a first-time flyer. Facing into the wind they commence their run, flapping their long unwieldy wings as fast as they can, while trying hard not to trip over their huge webbed feet. Sometimes they have to abort their takeoff at the last moment, tumbling into an undignified heap of wings and feet at the end of the runway. But once airborne they are transformed into the most graceful of fliers, soaring and wheeling on unmoving wings with all the grace of a ballerina. Their high landing speeds rob them of maneuverability, and their return to terra firma can be as undignified as their departure. Condors, in contrast, with the lower wing loadings of their low-aspect-ratio wings, have much lower landing speeds and can usually alight upon the ground without incident. Their possession of slotted wing tips – universally absent from the slender-winged marine soarers – may help to smooth their landings. Although slotted wing tips improve performance in vertical and in slow flight, they reduce efficiency in level flight (Drovetski 1996), which may explain their absence from marine birds, with their efficient high-aspect-ratio wings.

Thermal soaring can take place over land only when there are significant differences in surface temperatures, which happens only during daytime under the sun's influence. Grounded at night, soaring birds must wait until the sun heats the land before they can take to the air. The first thermals of the day are weak and can support only those birds with the lowest wing loadings – the smallest ones. This was clearly demonstrated by the meticulous observations of E. H. Hankin, who wrote an outstanding book on bird flight at the beginning of the century. Hankin lived near a meat processing plant in India, where the smells attracted carrion-feeding birds from far and wide. He went to great lengths to record the birds' flying performances, as well as their aspect ratios and wing loadings. He observed that as the early morning sun strengthened, and thermals formed, the smallest soarers flapped their wings and took to the skies. These were followed by larger birds, until finally the largest ones were able to soar. Landings at day's end were repeated in reverse order; the smallest birds were the last to be grounded by the failing thermal systems.

Large birds are obliged to soar because of the excessive energy demands of flapping flight. However, it is not their prerogative alone, and animals as small as butterflies take advantage of soaring when conditions are favorable.

PRACTICAL

In the first part of the practical you will be conducting some experiments on airfoil models using a wind tunnel. There are two models: one is a simple wing with a solid airfoil cross section, the other is hollow and is perforated by a series of holes on its upper and lower surfaces. These holes can be attached to manometers using lengths of thin rubber tubes. The remainder of the practical will be spent looking at some avian specimens and carrying out a simple experiment on a feather.

PART 1

Set up the balance for measuring vertical forces, with the pointer set vertically as described on page 204 and shown in Figure 12.12. Both airfoil models are provided with pointers that can be attached to their supporting rods

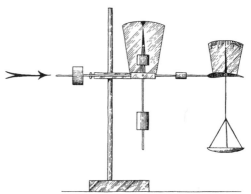

Figure 12.12. The apparatus used to measure lift generated by a model wing.

by metal collars, each fitted with a tightening screw. Before starting any of the experiments, attach the pointers to the airfoils. To do so, lay each airfoil on a flat surface and attach the pointer so that it is vertical, using a set square for reference. The balance arm of the apparatus is equipped with a graduated scale, marked off in degrees, so when the airfoil is attached, you can determine its angle of attack by referring to its attached pointer.

Experiment 1   Measuring the lift generated by an airfoil section at different airspeeds, set at a zero angle of attack.

Attach the airfoil model to the balance by inserting the supporting rod into the locating hole. Set the angle of attack to zero and tighten the screw. Adjust the position of the apparatus so that the airfoil, which is set to face into the airstream, has its horizontal axis coincident with that of the wind tunnel (Fig. 12.12). Level the balance so that the large vertical pointer is on the zero mark, and turn on the wind tunnel at its lowest speed setting. The lift generated by the airfoil raises the right-hand side of the balance, causing the vertical pointer to swing away from the direction of flow. This, in turn, causes the airfoil to assume an obtuse angle of attack. As we are measuring lift at a zero angle of attack, it is necessary to restore the beam to the horizontal position *before* adding weights to the weight pan. Do this by pulling gently on the pan. Once the beam is horizontal, add weights until you have added enough to equal the lift from the airfoil. At this point the balance will remain horizontal and the large vertical pointer will remain on the zero mark when you remove your hand from the weight pan. Remember, we are trying to assess the weight that the airfoil will *just* lift at a zero angle of attack. Be patient; this determination is a bit tricky. Remember also that we are working to the nearest 0.5 g.

Repeat the procedure for the rest of the speed settings, tabulating your results. The airfoil may experience buffeting, but this cannot be avoided. You will find that the airspeed has to be quite high before appreciable lift forces are generated. Can you suggest why?

Sample Observations:

Airspeed (m/s)	Lift (g)
3.9	<0.5
4.5	0.5
5.1	1.0
6.0	1.0
7.0	2.0
8.0	3.0
10.2	5.0
12.0	8.0
14.4	11.5

As you can see from these results, the lift remains small until the airspeed is quite fast. The most probable reason is that the airfoil has a fairly small camber. It is therefore a high-speed, rather than a low-speed, wing section.

Plot a graph of the log of lift (y axis) against the log of the airspeed (x axis) and measure the gradient. You should find that it is close to 2, showing that lift (like drag) is proportional to the square of the airspeed (at high Reynolds numbers).

Experiment 2 Measuring the lift generated by an airfoil section at a constant airspeed, for different angles of attack.

With the angle of attack still set at zero, adjust the airspeed to just over halfway (8.0 m/s) and measure the lift, using the same procedure as in the first experiment. By adjusting the angle of attack in 2° increments, measure the lift generated by the airfoil until it stalls (at an angle of attack of about 20°).

You will find that the rate of increase in lift tapers off toward the higher angles of attack. The drag increases very rapidly with increasing angles of attack, but we cannot measure it. If we could, we would find that the lift-to-drag ratio dropped off at high angles of attack. Stalling will probably be indicated by buffeting and by a decline in lift. Make a note of the approximate stalling angle. Plot a graph of lift (y axis) against angle of attack (x axis).

Sample Observations:

Angle of attack (degrees)	Lift (g)
0	4.0
2	6.5
4	12.0
6	15.5
8	19.0
10	21.5
12	25.0
14	27.5
16	29.5
18	30.5
20	approx. 24 (buffeting)

Experiment 3 Investigating the changes in air pressure over the upper and lower surfaces of an airfoil section at varying airspeeds, at a zero angle of attack.

You are provided with a set of manometers that are connected by fine rubber tubing to a series of small holes on the surface of an airfoil (Fig. 12.13A, B). Manometers 1, 3, 5, and 7 communicate with the upper surface; 2, 4, and 6 communicate with the lower surface (Fig. 12.13C). At the start of the experiment you should adjust the moveable manometer arm until the liquid levels in all of the manometers are level with the horizontal line. With the airfoil section set at a zero angle of attack, and the rheostat setting at the slowest airspeed, start up.

Increase the airspeeds one notch at time and notice the changes in pressure. You will notice that the largest pressure differences occur at about the shoulder (at about hole #3), and that the greatest differences are due to reduced pressure on

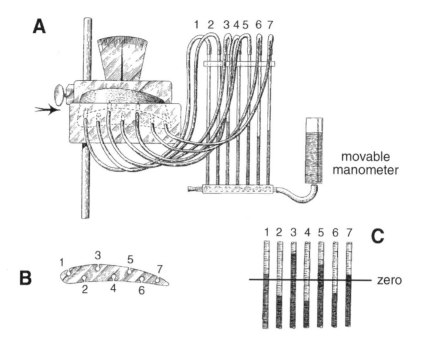

Figure 12.13. The apparatus used to determine the pressure changes over the upper and lower surfaces of an airfoil: (A) the complete setup; (B) how the seven holes relate to the two surfaces of the airfoil; (C) the pressure differentials in the seven manometers, with respect to atmospheric pressure (zero).

the top surface rather than to elevated pressure on the lower surface. When the wind tunnel is at full speed, measure the liquid levels in the seven manometers (Fig. 12.13C). Record your results in millimeters and note whether they are higher or lower than the atmospheric pressure (above or below the horizontal).

Experiment 4   Investigating the changes in air pressure over the upper and lower surfaces of an airfoil section at a constant airspeed, for varying angles of attack.

The same apparatus is used as in the previous experiment. Before commencing, attach a row of four or five small rectangles of tissue paper (about 10 mm × 4 mm) to the top surface of the airfoil using small pieces of Scotch tape attached to their leading edges. They should be evenly spaced and staggered along a straight line that lies parallel with the chord, approximately equidistant from either end (Fig. 12.14). If you have access to Mylar (a very thin plastic film, somewhat like the sort used in the kitchen), use 10 mm × 1 mm strips of this material for really good results. When the air flows over the airfoil these tabs will lie flat against the surface. However, when separation begins, they will pop up.

Set the airspeed at the highest setting (14.4 m/s) and measure the pressures at angles of attack of 0, 5, 10, 15, and 20°. Keep increasing the angle of attack, beyond the scale if necessary, and note the point at which the pressures suddenly drop off. At this point the airfoil has stalled.

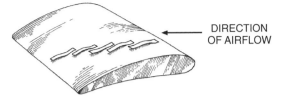

Figure 12.14. A model wing with small strips of paper attached.

DIRECTION OF AIRFLOW

At small angles of attack, the airflow maintains contact over most of the airfoil's surface. With increasing angles of attack the airflow begins separating from the surface. Separation commences posteriorly and moves further forward with increasing angle. Separation is complete at the stalling point. Notice how the pieces of paper attached to the airfoil flap violently when the airfoil stalls.

When an aircraft wing stalls it generates only a fraction of its maximum lift – less than the mass of the aircraft. There is also a great increase in drag, all of which makes flight unsustainable.

If you have the stomach for it (if you are a smoker this will be no problem), there is a nice experiment you can do with this apparatus to visualize the airflow over the airfoil. I am not a smoker, and I do not like puffing on cigarettes, but I survived this experiment. With the Surgeon General's health warning in mind, proceed, if you care to, with the next experiment.

Experiment 5 Examining the airflow over an airfoil section at varying speeds and at varying angles of attack.

Disconnect the rubber tubes from the manometers, but keep the tubes attached to the airfoil. With the angle of attack set at zero and the speed at the minimum setting, switch on the wind tunnel. Light a cigarette and slowly puff a mouthful of smoke into the tube leading to hole #1 (the most anterior hole in the upper surface). Make sure your eye is as close as possible to the level of the top surface – that way you will see the smoke stream at its best. Repeat for holes 3, 5, and 7 (all on the top surface).

Notice that the smoke is deflected back along the surface of the airfoil, as you would expect. Notice how close the smoke lies to the surface. Remember that the lowermost part of the boundary layer is stationary, and that the fluid flow becomes increasingly more rapid with distance from the surface. The boundary layer separates the smoke from contact with the surface, so the closeness of the smoke to the surface of the airfoil gives an idea of the thickness of the boundary layer.

To save you unnecessary suffering, select hole #3, which is at the shoulder. Puff smoke through #3, making a mental note of how close the smoke stream lies to the surface. Keeping your rate of smoke blowing constant, increase the airspeed to maximum. Do you notice how the smoke stream now lies closer to the surface, showing how the boundary layer becomes thinner with increasing Reynolds numbers?

Keeping the airspeed at maximum, blow smoke through the most posterior

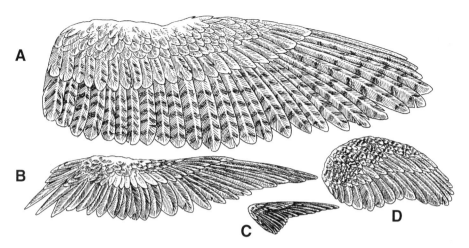

Figure 12.15. The wings of selected birds: (A) great horned owl (*Bubo virginianus*); (B) mallard (*Anas platyrhynchos*); (C) barn swallow (*Hirundo rustica*); and (D) the American woodcock (*Scolopax minor*).

hole (#7) while increasing the angle of attack. A point will be reached when the smoke stream breaks away from the surface – the airflow has separated from the airfoil. Gradually reduce the angle of attack until you find the minimum angle where separation occurs (about 25°–30°). Keeping the airfoil at this angle of attack, try blowing smoke through holes 5, 3, and 1, in turn. You will find that the airflow has not separated from the more anterior region of the airfoil. Indeed, you will have to increase the angle of attack considerably before you can achieve separation at hole #1.

For the last part of the experiment, set the speed to the minimum and the angle of attack to zero. While blowing smoke into the shoulder region (#3), increase the angle of attack until the airflow separates – the airfoil has stalled. Now gradually increase the speed setting to the maximum. Notice how the airflow over the surface is restored. Stalling can be corrected by increasing airspeed. Stalling occurs when an aircraft is flying with too high an angle of attack, or at too low a speed. A pilot gets some warning of the stall when the aircraft starts buffeting. At this point the pilot knows that the speed has to be increased or the angle of attack decreased. Both can be accomplished by pitching the nose down.

PART 2

You are provided with wing specimens of the following birds: great horned owl, mallard, barn swallow, and woodcock (Fig. 12.15). They are delicate, so handle with care. Compare these four wings, taking particular note of their aspect ratio and camber.

The owl, being a bird of prey, has to have a wing that can lift relatively large loads. Accordingly, the wing has a high camber. If you run your hand over the wing, it feels downy soft, and the feathers glide past one another smoothly and silently. If you try this with one of the other wings, you will find that they feel quite hard, and the overlapping feathers rustle past one another like stiff paper. The reason for the softness of the owl's wing is that the top surface of each feather is covered by a fine down of minute, soft bristles. This soft layer silences the movements of the feathers as they slip past one another when the wings flap. Another striking feature is a prominent fringe along the leading edge of the foremost wing feather. It is thought that this fringe smooths the airflow over the leading edge of the wing, preventing turbulence and thereby reducing noise. Most owls rely on their keen sense of hearing when hunting, hence the evolution of silent wings.

The woodcock, like other game birds, is adapted for rapid takeoff. Notice the low-aspect-ratio wing. Although good at fast takeoffs, the wood-cock is unable to sustain the pace for very long – low-aspect-ratio wings are good for acceleration but poor for endurance flying.

The mallard has a fairly high-aspect-ratio wing, giving it a high lift-to-drag ratio. Ducks are good strong fliers that frequently make long-distance migratory flights and therefore need efficient wings.

The aspect ratio of the wing of the common barn swallow is consider-ably higher. Swallows are among the fastest of all birds. Notice, too, the wing's slight camber and thinness, strategies for minimizing drag. High-speed aircraft have similar features.

Experiment 1   Assessing the airfoil properties of a contour feather.

Lay a feather on the table right-side up (convex surface up). Attach the quill to the table with a small piece of tape so the feather hinges freely. With the feather flat on the table, line yourself up parallel with the surface of the table and gently blow air over the feather. The lift generated causes the feather to rise.

Experiment 2   Making a flying wing from a sheet of paper.

You are provided with a sheet of paper, some thread, some Scotch tape, and a hair dryer.

Cut a strip, 10 cm wide, from a regular sheet of paper, making sure you get a straight edge. Fold the strip in half, flattening down the crease well. Fold one of the facing edges back upon itself to form a narrow lip, no more than 5 mm wide (Fig. 12.16). The other half of the strip will now be slightly wider than the one with the lip. Scotch tape two lengths of thread to the inside of the crease, in line with it. Tuck the wider half of the strip into the lip, so that it forms the convex surface of your paper wing. If necessary, hold the lip flat against the curved surface with some Scotch tape. Tether the wing to the top of a table, as shown in the figure, using the threads and some more Scotch tape. Holding a hair dryer in line with the wing and

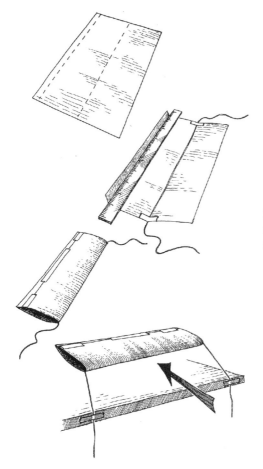

Figure 12.16. Making a flying wing model from a sheet of paper. The arrow shows the direction of airflow.

at least 15 cm from the edge of the table, blow a stream of air over the wing (use the slowest speed). The wing should lift off and hover pleasingly above the tabletop. If the wing experiences violent buffeting, try moving the hair dryer farther away. If that does not correct the problem, lower the dryer below the table and then, while keeping the nozzle parallel, raise it slowly above the level of the table until the wing takes off.

Swimming and Swimmers

got a really special present last Christmas from one of my old graduate students. I have it beside my computer right now, but I had better not start playing with it, otherwise I will never get this chapter written. The object of my attention is a small model of a shark, made from a particularly flexible rubber. It comes attached to its own lead, a length of flexible multicore wire, inserted into its middorsal line, just anterior to the pectoral fins. It is called a Twiddlefish[1] because when you twiddle the flexible drive, the model waggles like a fish. And when you put the model into water and start twiddling, it comes to life. Indeed, its swimming movements are so realistic that you would feel sure you had a real live fish at the end of the line. Twiddling the flexible drive sends waves of undulation down the body from the head to the tail, where the displacement is greatest. And this movement is just how most fishes generate their propulsive thrust of swimming.

Fishes, with over 30,000 known species, are the most abundant of living vertebrates, and their early origins can be traced back to the Devonian Period, over 360 million years ago. They therefore have a long history of swimming, and their undulatory mode of progression, so elegantly performed by my Twiddlefish, is probably the most primitive form of locomotion among the vertebrates. Not all fishes, far less all swimmers, use body undulations to generate their forward propulsion, but before looking at these alternate mechanisms, I will discuss the primal one.

Undulatory Swimming

If you enjoy eating fish you may have noticed that when you remove the skin you find a series of parallel zigzag lines running down both sides of the body, at right angles to its longitudinal axis. These thin white lines are of connective tissue – mostly collagen – and are the edges of the thin partitions, called *myocommata*, that separate the successive myotomes. The *myotomes* are the muscle blocks that all vertebrates have, which are arranged segmentally along the body in register with the vertebrae. If you examined the skinned fish with a lens you would see that the muscle fibers in the myotomes were attached at either end to the myocommata. The fibers are consequently short, and they lie parallel to the longitudinal axis of the body (Fig. 13.1). If your skinned fish is cooked and you try flaking the myotomes apart, you will

myocommata

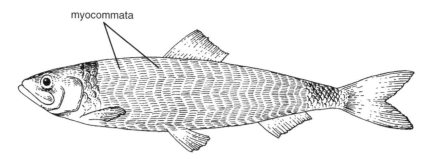

Figure 13.1. A skinned fish, showing the underlying zigzag pattern of myocommata between adjacent blocks of body muscles.

notice that they fit together with a complex cone-within-cone structure. Functionally, this arrangement causes a wave of undulation to pass down either side of the body when the muscle fibers contract.

The wave of contraction commences at the anterior end of the trunk and passes down toward the tail (Fig. 13.2). As the posterior part of a fish's body is thinner, and therefore more flexible, than the anterior portion, the amplitude of the wave is greatest at the tail end. Just how these waves of undulation produce a forward thrust can be explained by a simple analysis of the components of the wave and some thought about the action and reaction of forces. The explanation that follows is probably a gross oversimplification of a complex mechanical problem, but I believe that the basic concept is sound.

Consider first the forces acting on the right side of the fish's body. The

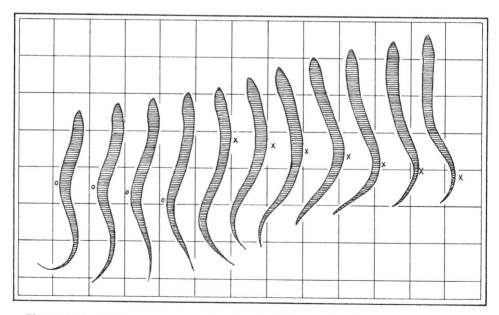

Figure 13.2. Time-lapse sequences of a swimming fish, seen from above. Waves of contraction pass down the left (circles) and right side (crosses) of the body.

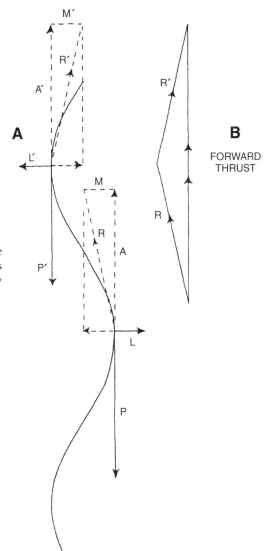

Figure 13.3. A simplification of the forces acting on a fish due to the waves of contraction passing down its body (A). The resultant is a forward thrust (B).

wave passing down the length of its body toward the tail has two components, a posterior one (P in Fig. 13.3) and a lateral one (Fig. 13.3, L). Water is a liquid, and its molecules are free to flow. However, it should be remembered that water, which is more than 800 times denser than air, offers resistance, too, as everyone knows who has ever made a belly-flop landing in a swimming pool. The posterior and lateral components of the wave therefore react against the resistance, or inertia, of the water. According to Newton's third law, actions and reactions are equal and opposite, so the fish experiences an anterior reaction against the posterior component and a medial reaction against the lateral one (Fig. 13.3, A and M, respectively). The resultant of the anterior and medial reactions is a thrust directed anteromedially (Fig. 13.3, R), pushing the fish forward and toward its left side. The same argument applies on the other side of the body, and this resultant is also directed anteromedially. However, since the resultant is acting on the left side of the body, its line of action is directed forward and toward the fish's right side. Since these two anteriorly directed resultants are inclined toward opposite sides of the fish, their sideways components cancel out (Fig. 13.3B) and the resultant is a forward thrust, though the fish still tends to yaw as it swims forward in a straight line.

Vertebrates that derive all of their forward thrust from body undulations, which include eels, aquatic snakes, crocodiles, and marine iguanas, are described as being *anguilliform,* meaning eel-like. Although anguilliform swimmers are in the minority among vertebrates, most fishes probably obtain at least some of their forward thrust from undulatory movements of the body, augmented to a lesser or greater extent by thrust from the tail. Many fishes, like those living among coral reefs, derive their thrust from movements of the median or paired fins, but we will not be concerned with them.

Tail Propulsion

Given that the posterior end of an animal experiences the greatest degree of displacement during body undulation, it follows that the tail region should have undergone early specialization as an organ of propulsion during vertebrate evolution. In the ichthyosaurs, for example, the earliest members of the group, like *Chensaurus* and *Utatsusaurus* from the Lower Triassic, had straight vertebral columns, with some evidence of a fleshy expansion in the caudal region (Motani, You, and McGowan 1996). By Middle Triassic times *Mixosaurus* showed some specialization in the caudal region, with a change in height and slope of the neural spines. This suggests the presence of some sort of tail structure. Furthermore, there is evidence that *Cymbospondylus,* also from the Middle Triassic, had a slight tailbend. *Shonisaurus,* from the Upper Triassic, also had at least a shallow tailbend (Kosch 1990; McGowan 1991). By Early Jurassic times the marked tailbend, characteristic of post-Triassic ichthyosaurs, was well developed and supported a large crescentic tail (Fig. 13.4). That the vertebral column evolved a downward kink, thereby supporting the lower lobe of the tail, rather than an upward one to support the upper lobe, may have been fortuitous rather than functional, and the selection pressure may have been simply to increase the area of the tail (McGowan 1992).

The tail, or more precisely, the caudal fin, comes in a wide variety of shape, size, and stiffness, even within a given group of vertebrates, such as within the sharks. The caudal fin of the spiny dogfish (*Squalus acanthias*), for example, has a low aspect ratio and is very flexible. The blue shark (*Prionace glauca*) has a more crescentic caudal fin, with a higher aspect ratio, but this too is flexible, like the body. The mako shark (*Isurus oxyrhynchus*), in contrast, has a remarkably stiff tail, in keeping with its much more rigid body, and this tail has a

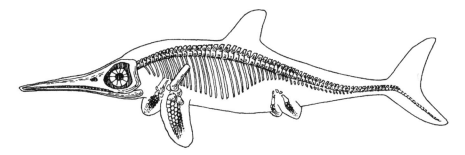

Figure 13.4. An Early Jurassic ichthyosaur.

high aspect ratio. Regardless of the diversity of form among caudal fins, most of them, except perhaps the most flexible ones, generate forward thrust by acting as inclined plates.

If we could look down on a swimming fish and watch its movements in slow motion, we would see that the caudal fin is deflected from side to side by drag forces. This is because of the flexibility of the *caudal peduncle,* the narrow region immediately anterior to the caudal fin. Thus, as the peduncle moves to the fish's right the trailing edge of the caudal fin lags behind the leading edge, and therefore it becomes an inclined plate, set at an acute angle of attack (Fig. 13.5). The lift generated by the plate is directed toward the

Figure 13.5. A swimming fish, seen from above, showing how the caudal fin acts as an inclined plate, generating a lift force directed toward the head.

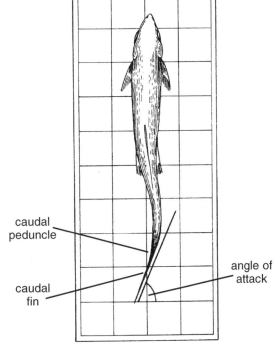

caudal peduncle

angle of attack

caudal fin

fish's head. When the caudal fin has completed its swing toward the right and starts moving back toward the fish's left, the caudal fin becomes deflected in the opposite direction. Again the trailing edge lags behind the leading edge, so that the caudal fin becomes an inclined plate set at an acute angle of attack, generating an upthrust toward the head. Each stroke of the tail is therefore a power stroke, propelling the fish forward in the water.

There is a continuous spectrum among swimming vertebrates in the degrees to which the caudal fin and body undulation contribute to propulsion. At one extreme are pelagic fishes like tunas and swordfishes (scombroids) and the cetaceans, collectively referred to as being *thunniform* (tunalike). Thunniform swimmers have stiff bodies, with lateral movements being confined to their stiff, high-aspect-ratio tails.

Cetaceans, it should be noted, do not have such high-aspect-ratio tails because, unlike those of scombroid fishes, they have no internal bony skeleton and are consequently not as stiff. Here is a good example of how the properties of materials, in this case connective tissue, determine the mechanical properties of the structure, namely the tail flukes. It is also an example of how phylogeny can place constraints on evolutionary opportunity. Cetaceans have been unable to evolve the more efficient high-aspect-ratio tails of scombroid fishes because there are no bones in the terminal region of the mammalian tail that can be used to stiffen the tail flukes. Lamnid sharks – the group that includes the great white shark – have thunniform bodies, but their caudal fins, like those of cetaceans, do not have such high aspect ratios as those of scombroid fishes. This is because the internal supporting skeleton is cartilaginous and therefore has a much lower Young's modulus than bone, and also because much of the caudal fin is skeletally unsupported.

Thunniform swimmers, by eliminating lateral movements of the body, greatly reduce their drag. Their crescentic tails have a high lift-to-drag ratio, and they also possess other drag-reducing features, which I will describe in a moment. At the other end of the spectrum are fishes like the salmon and cod. Their low-aspect-ratio tails are broadly fanned, and they obtain much of their propulsive thrust from undulating the body. Such fishes are referred to as being carangiform. Their swimming movements generate considerably more drag, and they are therefore nowhere near as efficient as thunniform swimmers. Indeed, fishes with flexible bodies generate perhaps as much as two to five times more drag than thunniform swimmers (Webb 1982).

To my thinking, the body of a tuna represents the highest achievements of marine engineering in the living world. The body is beautifully streamlined, and if you ran your hands over its surface it would feel as smooth as the finish on a luxury car (Fig. 13.6). To minimize drag the fins are thin, and whereas those of the swordfish project stiffly from the body, most of the tuna's fins can be tucked out of the way when not being used for changing directions. The anterior dorsal fin can be completely collapsed into a groove, as can the paired pelvic fins. The pectoral fins are flattened against the side of the body, and although they are

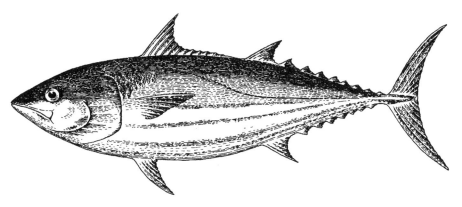

Figure 13.6. The tuna, with its beautifully streamlined body and narrow caudal fin, is adapted for cruising.

thin, the body is recessed for their reception so that they lie absolutely flush with the surface. The free margin of the operculum, the flap that covers the gills, similarly fits into a shallow recess so that it, too, lies perfectly flat against the body. Even the eyeball is contoured to the body surface, and if you closed your eyes and ran your fingers over the smooth head, you would feel little discontinuity as they traced over the tuna's eyes. The significance of having a smooth body is that the major drag component in a streamlined body is friction drag (Chapter 11), which is minimized by the smooth surface.

Aside from being more than 800 times denser than air, water is also more viscous, by a factor of about 55 at 20° C. As Reynolds number is an expression of the ratio of density to viscosity (Re = lsd/v), it follows that objects moving in water have Reynolds numbers that are about fifteen times higher than they would be in air, other things being equal. Tunas are large fishes, the largest species reaching lengths of up to 4.3 m (14 feet), so they swim at high Reynolds numbers. As Reynolds numbers increase, the boundary layer not only becomes turbulent but also begins separating from the surface of the body. Separation begins posteriorly and moves forward, as we saw in the experiments with airfoils. It can be delayed by causing the flow to become turbulent, as in dimpled golf balls, because a turbulent boundary layer adheres more closely to the surface than a laminar one. It has therefore been suggested that the tuna's median finlets, which lie along the dorsal and ventral edges of the body just anterior to the caudal fin, may function as vortex generators, like those at the root of the tail of the Boeing 737 (see Wardle 1977 and the references therein).

Cruising, Accelerating, and Maneuvering

Thunniform swimmers, like tunas, are specialized for cruising, their stiff bodies and high-aspect-ratio tails minimizing their transport costs. Specialization in one direction always entails compromises in others, and thunniform

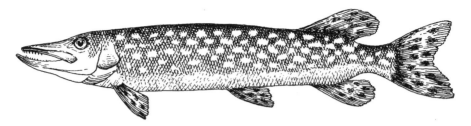

Figure 13.7. The pike, which is adapted for sprinting, has a broad tail whose area is effectively increased by the caudally placed anal and dorsal fins.

swimmers have very poor acceleration. This can be demonstrated for some species by catching and releasing them, whereupon their tails oscillate furiously without their making any significant headway until the tails eventually "bite" – something like a car spinning its wheels from a standing start. The reason for their lack of acceleration is that the surface area of their caudal fin is relatively small, and the aspect ratio is high.

Fishes specialized for rapid acceleration include the pike (*Esox*), a sit-and-wait predator that lurks in the vegetation until a suitable prey comes its way (Fig. 13.7). It then lunges forward, snapping the prey up with its well-armed jaws. Several features contribute to its rapid acceleration. The caudal fin is large, with a low aspect ratio. The median dorsal and anal fins are also large and are set so far back that they are essentially a functional part of the caudal fin, too. Consequently, when the pike flicks its tail, a large volume of water is displaced, generating a large forward thrust. Furthermore, the body is long and deep, presenting a large surface area against the water and maximizing the propulsive thrust of the body undulations. The cost of rapid acceleration is high drag, and lungers like the pike are inefficient swimmers.

The third major category of fishes, as so clearly explained by Webb (1984), are those specialized for maneuvering. Included here are all those deep-bodied forms, like the butterfly fish (*Chaetodon*), that live among the corals. They lack the pike's acceleration and the tuna's high-efficiency cruising, but when it comes to maneuvering in and out of the small spaces between corals, they have few equals. Their great maneuverability is partly attributed to their ability to undulate their median fins, and these waves of undulation are reversible, so the fishes can move forward or backward with equal ease.

Although many fishes can be placed into one of these three locomotory categories, most fishes are generalists, being able to perform moderately well in all three.

Stability and Control

Bodies that move in three dimensions, whether they are fliers or swimmers, require both stability and control. Stability is the automatic correction that occurs when a moving body deviates from a straight course, as when the

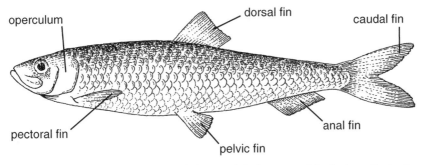

Figure 13.8. Bony fishes primitively have pelvic fins set well back along the body and ventrally placed pectoral fins.

flight feathers of a dart remove any wobble. Control, in contrast, is the action a body takes to change course. Both functions are performed by the fins.

Fishes have paired pectoral and pelvic fins and unpaired median dorsal and anal fins (Fig. 13.8). There may be two dorsal fins, and sometimes the dorsal and anal fins are elongate and may be undulatory, providing for propulsion. The pectoral fins are primitively placed low on the body, behind the level of the operculum, and the pelvics are placed well back, often just posterior to the level of the deepest part of the body. However, in many more derived fishes, the pectoral fins are placed high on the side of the body and the pelvics lie well forward, often on a level with, or even in front of, the pectorals. Some fishes, such as the swordfish, have lost their pelvic fins. The following tentative comments on stability and control will be based upon more primitive fishes, like the herring.

Yaw is probably largely corrected by the anal fin, in the same way that an airplane's vertical rudder acts as an inclined plate to eliminate yaw. The dorsal fin probably has the same role in those fishes where it is placed behind the center of mass. Yaw might be induced by the extension of one of the pectoral fins, which would cause the fish to slew around to that side. Alternatively, it may be brought about by the lateral flexion of the body.

Pitch is probably corrected for passively by the pelvic fins, acting like the horizontal stabilizers of an airplane. The pectorals, being placed in front of the center of mass, would only make pitching movements worse, but they are effective at causing pitching, acting like the hydroplanes of a submarine. Inclining the pectoral fins upward at an acute angle of attack would generate an upthrust, pitching the head up, whereas the reverse would be achieved by tilting their leading edge below the horizontal.

Roll is probably corrected by the damping effect of the dorsal fin "biting" into the water as the fish rolls to one side. Roll might be effected by generating differential lift forces in the pectoral fins, either by partially extending one of them or by setting them at different angles of attack.

Stability and control are inversely related; a body that is especially stable is more difficult to turn than one that is not so stable. Thus, an airplane with a particularly large tail, or a fish with a large dorsal fin set well behind the center of mass, would need relatively large forces to change its course. Some fishes have dorsal fins set well forward along the body, partly or wholly anterior to the center of mass. Such anteriorly placed dorsals are destabilizing, because when the fish turns, the fin acts as an inclined plate that generates a thrust in the direction of the turn. This therefore enables a fish to turn more readily, just as a person sitting at the front of a canoe has far more steerage control than a person sitting at the stern.

Swim Bladders

Most bony fishes have a swim bladder, a gas-filled chamber that lies just below the vertebral column. Although some open into the gut, most swim bladders are closed, and their lining secretes and absorbs gases. Fishes are therefore able to maintain the volume of the chamber, thereby matching their density to that of the surrounding water regardless of their swimming depth and maintaining neutral buoyancy. This feature explains how the fishes in an aquarium are able to remain motionless in the water, neither rising nor sinking. When fishes are caught in a trawl net and brought to the surface, their swim bladders expand, due to the decrease in pressure. Many fishes are unable to cope with the rapid change, and they come to the surface with their distended swim bladders bulging from their mouths. Cartilaginous fishes lack a swim bladder.

Fish Slime and Drag Reduction

Fishes are slippery to the touch because they produce mucus that is discharged onto the skin. Mucus has several functions, including acting as a barrier to water[2] and providing protection against parasites and abrasions. Another function of mucus, used by many species but apparently not by tunas and their allies, is drag reduction (Hoyt 1975). The active agent is a water-soluble polymer – actually there are many of them – that functions even at low concentrations. One of the most effective of these polymers is produced by the Pacific barracuda, a fast predator. Experiments have shown that a 5 percent solution in seawater reduces drag by 65 percent. The United States Navy conducted research on fish mucus, doubtless to improve the performance of submarines. However, since the polymers rapidly deteriorate after removal from the fish, losing their effectiveness in a matter of hours, they have no practical application for the Navy.

Gray's Paradox

In 1936 Sir James Gray, a leading light in the field of animal locomotion, published a paper on the swimming performance of a porpoise that posed a problem so baffling that it became known as "Gray's paradox." The gist of the problem was that a porpoise was able to achieve a speed of 37 km/h (23 mph) for 7 seconds, a feat seemingly possible only if it reduced the drag on its body to a fraction of the predicted value. With the advent of the Cold War in the 1950s, interest in Gray's paradox shifted from biological circles to military ones, and the United States Navy began investigations to find out whether the drag-reducing mechanisms of cetaceans could be applied to submarines. The first step was to gather accurate data on the swimming performance of dolphins and porpoises, and this research began in 1960 at the Naval Ordnance Test Station in California. Conventional submarines of that era operated at Reynolds numbers as high as 10^9, under which conditions flow is probably turbulent throughout the entire boundary layer. Cetaceans, in contrast, were thought capable of maintaining laminar flow throughout much of the boundary layer. If such a degree of laminar flow could be achieved for submarines, drag would be drastically reduced and speed would be correspondingly increased for the same engine power. A clue to the drag-reducing mechanism of cetaceans appeared to lie in their skin – its spongy texture possibly damped disturbances in the boundary layer before it became turbulent. Experiments were conducted with various resilient coatings, but no reductions in drag were achieved.

As more data were gathered it began to dawn on the researchers that cetaceans did not possess any special drag-reducing mechanisms at all. Indeed, their transient bursts of high speeds could be achieved with the boundary layer completely turbulent, as in submarines. Their only "secret" was that their bursts of power were more than three times higher than Gray had calculated. Gray's estimate for his porpoise's power output was based on what a human athlete could maintain for a 15-minute period, whereas the cetacean's burst lasted only 7 seconds. The porpoise's power output was therefore grossly underestimated during its short burst; hence the drag forces on its body were underestimated, too. In hindsight we can see where Gray went wrong, but his was a reasonable premise at the time, and it did stimulate a great deal of research on cetacean swimming.

Swimming Speeds of Large Pelagics

Many of the large pelagic fishes, like the swordfish and its allies, appear to be built for speed, and there are various accounts of game fishes like marlins reaching speeds of 130 km/h (80 mph). But fishermen's stories are to be treated with caution, and the serious studies that have been conducted have recorded speeds in the region of only a few kilometers an hour. Blue marlins (*Makaira nigricans*) tagged with electronic speed-recording devices, for example, spent more than 97 percent of their time swimming at speeds of less than 4.4 km/h (2.7 mph), and their top speed was only 8 km/h (5 mph), which seems remarkably slow (Block and Booth 1992). In another study, five swordfishes were tagged with radio transmitters and monitored for five days (Carey and Robinson 1981). Their estimated swimming speeds did not exceed 3.6 km/h (2.2 mph). The same techniques gave estimated sustained swimming speeds for bluefin tuna (*Thunnus thynnus*) of 5.4 km/hr (3.3 mph), which accords with experimental results obtained within a flume (Dewar and Graham 1994).

What should we make of these seemingly fast pelagic fishes that appear to spend most of their time cruising along at only a few kilometers per hour? Drag forces are high in water because of its high density, and these forces increase with the square of the speed at high Reynolds numbers. So if a swordfish accelerated from 2 to 8 mph, the drag force would increase by a factor of 16. Here is reason enough for putting a limit on cruising speeds. But another aspect of drag has to be taken into account, namely, the phenomenon of critical Reynolds numbers.

Briefly stated, as Reynolds numbers rise, a critical point is reached where the flow in the boundary layer changes from being laminar to turbulent, causing a sharp increase in drag. The critical Reynolds number at which this transition occurs depends on the shape of the body and is probably in the order of 10^6 for the fishes we have been considering. Keeping speeds low maintains low Reynolds numbers, and it may be that large pelagic fishes swim at low cruising speeds to keep from reaching their critical Reynolds numbers, thereby avoiding the increased drag accompanying a turbulent boundary layer (Wardle 1977).

The Largest Vertebrates

Since water is 800 times denser than air, it follows that objects immersed in water are correspondingly more buoyant than they are in air. This is because the upthrust on an object is equal to the weight of fluid displaced (Archimedes' principle). The buoyancy of air is negligible, except for lighter-than-air objects like helium balloons, but it is considerable for water. Most animals have a density close to that of water, so aquatic animals are weightless when submerged and consequently can grow to enormous sizes. It is no coincidence that the largest living animal, the blue whale (*Balaenoptera musculus*), is aquatic. Reaching lengths in excess of 30 m (100 feet) and masses of over 200 tons, it is the largest animal that has ever lived, heavier even than the largest sauropod dinosaurs.

PRACTICAL

OBSERVATIONS ON FISHES

If you have access to an ichthyology collection you should make some comparisons between the following fishes.

1. *An eel.* Notice the great length and flexibility of the body and the long dorsal and anal fins.
2. *A requiem shark.* This group includes the genus *Carcharhinus* and the

blue shark, *Prionace glauca*. Notice the flexibility of the body and the general lack of stiffness in the caudal fin.

3. *A lamnid shark.* This group includes the porbeagle (*Lamna nasus*), mako (*Isurus oxyrhynchus*), and the great white (*Carcharodon carcharias*). Notice that the body is far less flexible than that of the requiem sharks, and that both lobes of the caudal fin are stiff.

4. *A tuna.* Notice the smooth lines of the streamlined body. Check to see how the anterior dorsal and pelvic fins collapse into a groove, and how the pectoral fins and the trailing edge of the operculum are recessed into shallow depressions on the side of the body. Run your fingers over the eye and along the length of the body to see how smooth it is.

5. *A pike or muskellunge, of the genus* **Esox.** Notice the broad caudal fin and how far back the dorsal and anal fins are placed.

If you do not have access to a suitable fish collection, you could purchase a mackerel and small trout inexpensively from a supermarket. The mackerel is a scombroid fish, like the tuna, and is essentially thunniform. The body is smooth and beautifully streamlined. Notice the stiff caudal fin. Try spreading it out – it has a fairly high aspect ratio. Notice, too, the small finlets anterior to the caudal fin. The trout does not have such a gently tapered streamlined body and is more flexible than the mackerel. Notice the broad, flexible caudal fin. Compare the relative areas of the caudal fin and the body. You can do so by tracing around their outlines on graph paper and counting up squares (remember to spread the tail before tracing around it). If you repeated this for the mackerel and compared the results, you would find that the latter had a relatively smaller fin area.

The following two experiments can best be done at home, using the bathtub. You will need: some plasticine; four drinking straws; a sewing needle and thread; two paper clips; and a bathtub filled with cold water.

Experiment 1 Comparing the differences in drag between a solid cylinder and a streamlined model of the same length and diameter when they are towed in water.

Take a piece of plasticine and roll it into a rod about 15 cm (4 inches) long and 2 cm (3/4 inch) in diameter. Cut the rod in half. Leave one of the halves as a solid cylinder; trim and shape the other into a streamlined model, making sure to maintain the same frontal area and length (Fig. 13.9). The cylindrical shape will be heavier than the streamlined one and needs to be trimmed so the two have the same mass. Do this by using one of the straws as a corer and hollowing out the cylinder until it is about the same mass as the streamlined model.

Break off the inner loop of one of the paper clips and embed it into the leading end of the streamlined shape. Repeat for the cylinder. Using the needle and thread, attach each plasticine model to the end of its own drinking straw. Make the towing thread about 3 cm (1 inch) long.

Try pulling each of the models through the water. Use the length of the tub and see how fast you can go. Was the streamlined shape easier to pull?

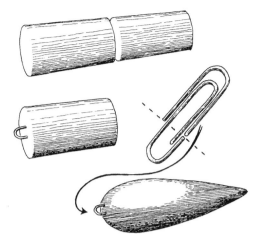

Figure 13.9. Constructing models for towing in water.

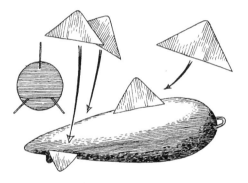

Even though the towing speed is low, the density of water is high, so the drag forces are consequently large. You would therefore have noticed a marked difference in drag forces between the two models.

Experiment 2   Assessing the stabilizing role of fins.

You will need an aluminium baking tray. Convert the cylindrical model from the previous experiment into a streamlined shape so that the two models are the same. Cut out three right-angled triangles from the aluminium tray. One of the triangles should have two sides of about 2 cm, the third side being the hypotenuse. The other two triangles should each have two sides of about 1.5 cm. Cut a shallow slit in the middle of the dorsal region of one of the models, the same length as the hypotenuse of the largest triangle. Fit the long edge of the large aluminium triangle into the slit, and smooth down the plasticine on either side. The triangle, which should be vertical and firmly anchored, is to simulate a dorsal fin.

Tow the unmodified streamlined model up and down the bath, inducing yaw and pitch by jiggling the towing straw. Also try making some turns. You should find that the model moves quite well, but that it is not very stable. Repeat with the modified model. You will probably find some improvements in its stability.

Following the same procedure as before, fit the two small triangles to the posterior end of the modified model to simulate pelvic fins. Incline them ventrally, at an angle of about 45°, like the paired fins of a swordfish. Return the model to the water and repeat the last trial. You will probably find a marked improvement in its stability. If you try inducing it to yaw and pitch, you will find that these movements are rapidly dampened out.

Chapter Notes and Responses to Questions

Prologue

1. Some theropods, including *Tyrannosaurus,* have only two fingers. This is a secondary loss.

Chapter 2

1. We tend to use freezing point for materials that are liquid or gaseous at room temperatures; we use melting point for those materials that are solid at room temperature.

Chapter 6

1. A Wheatstone Bridge circuit is simply an electrical circuit in which an unknown resistance (in this case the strain gauge) is compared with a known resistance, via a variable resistance. The variable resistance is adjusted until the known and unknown resistances are in balance, and no electricity flows in the circuit. From the reading of the variable resistance and the knowledge of the fixed one, you can deduce the unknown resistance. The circuit is explained in detail in all good high school physics texts.

Chapter 7

Observation of the Vertebral Column of the Horse.

The reason why the transverse processes are horizontal in the horse is because the back is kept essentially straight and is not used in a springing action during locomotion. The reverse is true in the dog, where the ventrally directed transverse processes provide a longer lever arm (see Chapter 9) for the muscles that flex the back.

1. There is some evidence that the giraffe may not have seven cervical vertebrae. See: Solounias, N. 1997. Remarkable new findings regarding the evolution of the giraffe neck. *Journal of Vertebrate Paleontology,* 17 (Supplement to number 3, Abstract of Papers), page 78A.
2. The mastodon in the gallery of the Royal Ontario Museum has a pathological condition: only one tusk is present, and the socket for the other tusk, which was never found, was much reduced in size.

Chapter 8

Friction Experiments

There is much experimental error, but you should have found that the limiting frictional force was independent of surface area. For example, I found that when the slab was placed face down, the area of contact was 12.5 cm², compared with only 6.2 cm² when placed edge down, but the limiting frictional forces were 20 g and 19 g, respectively.

You should have found that the limiting frictional force, F, increased with the loading force, W. For example, I found that when the cube was loaded with 10 g, 20 g, 30 g, and 50 g, the corresponding limiting frictional forces were 17 g, 25 g, 28 g, and 41 g, respectively.

Examination of Anatomical Specimens

Regarding the examination of the joint between the proximal end of the first metacarpal and the carpus in humans, the other place where you have seen saddle-shaped articular surfaces is in the joints between the vertebral centra of birds.

Chapter 9

Experiment 2

If the articulated arm skeleton is positioned such that the forearm is perpendicular to the humerus, the distance between the center of the insertion for the biceps and the middle of the fulcrum (taken as the middle of the articular surface at the distal end of the humerus) is about 60 mm. This is the moment arm of the effort. The distance between the fulcrum and the middle of the hand is about 320 mm. This is the moment arm of the load. The mechanical advantage is therefore 60/320 = 0.19. Therefore, the force exerted by the biceps muscle = load/0.19. Hence the muscle has to develop a force about five times greater than that of the load.

Chapter 10

1. Although the term *reptile* is not used formally by taxonomists, it is still in wide general usage and is a useful category in which to discuss locomotion.
2. The information about runners was provided by A. J. Higgins, of the Department of Athletics and Recreation, University of Toronto. He points out that children, being more supple in their limbs than adults, naturally tuck in their lower legs when they run.
3. A distinction is made here between 1 C° and 1° C, which are not the same things. The former is one division on the temperature scale, whereas the latter is a specific temperature (a little above the freezing point of water).

Practical, Experiment 8

The assumptions made in this experiment are as follows:

1. The hand holding the chalk does not move vertically for the duration.
2. The vertical displacement is entirely due to that of the center of mass and is not influenced by postural changes accompanying locomotion.

3. No energy is stored in the body during the stride cycle that can be used for elevating the center of mass. (This assumption is probably not true.)

For useful discussions of the energetics of locomotion and the storage of strain energy, see Alexander (1984b, 1984c).

Chapter 11

1. The frontal area of the towed line is more than just its cross-sectional area: you must also take into account the curvature as it dips into the sea and any snaking as it moves through the water.

Practical, Part 2, Experiment 2

The sphere consistently reaches the bottom before the streamlined shape, showing that it generates the lowest drag. This is contrary to intuition because the streamlined shape should have the lowest drag. However, the streamlined shape is a device for reducing pressure drag, which is significant only at high Reynolds numbers. Because of the short distance over which these models fall, and their low density, they do not reach high speeds. Furthermore, they are both fairly small. Consequently, the Reynolds number for both models is low (remember that $Re = L \times S \times D/V$). At a low Reynolds number, friction drag predominates, and this increases with greater surface area. Since the streamlined shape has the largest surface area, it follows that it experiences the largest drag forces.

Part 2, Experiment 3

The results are the reverse of Experiment 2 because the models are moving at higher Reynolds numbers. This is attributed to the greater distance traveled and to their larger size. Under these circumstances, pressure drag predominates, so the drag on the streamlined body is less than that of the sphere, as would be predicted.

Part 2, Experiment 4

The high viscosity of the glycerine slows the descent of both models, reducing the speed component of the Reynolds number equation and increasing the viscosity term. As a consequence, the Reynolds number of the two models is low, and the streamlined shape therefore experiences the greater drag force, as in Experiment 2.

Part 2, Observation 2

As the seahorse has a nonstreamlined body, it is reasonable to conclude that it does not swim at high Reynolds numbers. These fishes, which hold their bodies vertically in the water, swim fairly slowly.

Part 3, Experiment 1

When the inclined plates on the paper airplane are set at a small angle of attack (that is, the trailing edge is lower than the leading edge), they generate an upthrust. This upthrust, acting behind the center of mass, causes the nose of the airplane to tilt downward in flight. The reverse happens when the angle of attack is an obtuse angle (that is, when the trailing edge is higher than the leading edge). When the plates are inclined in opposite directions, the airplane spins about its longitudinal axis.

Part 3, Experiment 2

You should have found that when the high-aspect-ratio plate was placed above the square one, and the two were moved through the water, they separated. This is because the rectangular plate has a higher lift-to-drag ratio than the square one.

When the two oblong plates are tested together, with the one with the tapered profile on top, they should have separated slightly when moved through the water with an acute angle of attack. This is because the plate with the tapered profile generates less drag and therefore has a higher lift-to-drag ratio.

Chapter 13

1. Twiddlefish is the trademark of Twidco, P.O. Box 542, Durham, NC 27702. You can look for these wonderful models at public aquariums, but if you have difficulty locating a supplier, I am sure that Twidco will be able to help you.
2. Freshwater fishes absorb water because the osmotic pressure of their body fluids is higher than that of their environment, whereas marine species lose water for the converse reasons.

Instructor's Notes

LABORATORY 1

Equipment

Two balances. One should read in grams, to two decimal places, up to 200 g – a torsion balance with a digital readout is ideal. The other does not have to be more accurate than to the nearest 10 g, but it needs to be able to weigh objects up to about 10 kg.

Calipers, measuring up to 150 mm, to one decimal place. Rulers, measuring in millimeters. A steel tape, measuring in millimeters.

Specimens

Cubes

A set of plaster cubes, varying in size from 1 cm up to about 20 cm. The smaller cubes (up to about 4 cm) can be made by making a wooden model, using that to form a silicone mold, and then pouring liquid plaster into the mold to make a cast. Larger cubes can be made by making a wooden box of the appropriate dimensions and using that as the mold for casting. Both methods are outlined here. The reason for using the latter method for the larger cubes is that large silicone molds are not stiff enough to maintain their shape when filled with liquid plaster. If you can find someone who has gone through art school, he or she would be able to make a set of cubes for you very easily because most art students have learned about molding and casting. If not, you can make them for yourself using the following procedures. The first step is making the molds.

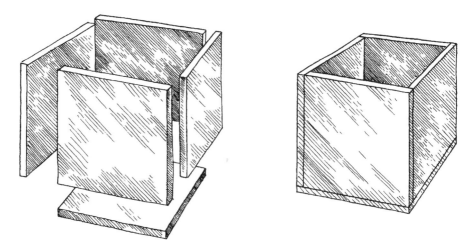

Figure A.1. Constructing a mold for making plaster cubes.

Small cubes Cut some pieces of wood to the required size, finishing off the faces, if necessary, by using a flat sheet of sandpaper. Varnish the wooden cubes, and allow them to dry. Purchase a tube of silicone bathroom sealant – this comes in a cardboard tube for use with a caulking gun, which you will also need. Spray the wooden cube with one of those nonstick oil sprays used in cooking. Alternatively, apply a thin film of oil with a rag or brush. The oil film acts as a separator, preventing the silicone from sticking to the cube when the sealant cures. Squeeze out some of the silicone sealant and, using a cheap paintbrush (it will be ruined by the silicone), apply a thin layer (about 2 mm thick) to all but one face of the cube. (Wrap the end of the nozzle of the dispenser with plastic wrap after each use to prevent the sealant from going hard and blocking the tube.) Allow the silicone to cure (probably an hour or so), and repeat. The second and subsequent coats can be thicker. If you have some cheesecloth (or gauze bandage), you can incorporate it into each layer to increase strength and stiffness. Continue until you have formed a sufficiently thick layer around the cube to retain its shape once the wooden model has been removed. The thickness depends on the size of the cube. For the smallest ones it need be only about 1 cm. Allow the silicone to cure for several hours before removing the wooden cube. Take care not to tear the silicone mold during this process – if you used cheesecloth this is unlikely to happen.

Large cubes You need to cut out five rectangles of 3/4-inch plywood, corresponding to five of the faces of the cube. These will be held together with panel pins (screws for the larger cubes) lapping the edges, so allow for the thickness of the wood when measuring the rectangles prior to cutting (Fig. A.1). Once you have made the box, varnish the inside surfaces and leave it to dry. Apply a thin film of oil, as in the previous case.

Making and pouring the plaster Plaster of Paris can be obtained from some building suppliers, hardware stores, and arts and craft shops. Alternatively, you may be able to obtain dental grade plaster from a dental supply company (check your local Yellow Pages or ask your dentist). Dental plaster, used in making casts of dentitions, is denser than the regular grade and is preferred.

Ideally you should have a series of molds, silicone and wooden, ready for pouring. Partly fill a plastic bucket with cold water – about 10 cm deep for the first batch. Add the plaster to the water, one handful at a time, by letting it sift between your fingers. The plaster will sink to the bottom, releasing bubbles of trapped air. Do not stir! Keep adding plaster until only a thin film of water remains. Now give the plaster a good stir. It should have the consistency of creamed soup. Resist the temptation to add more plaster – it is meant to be fairly thin. Gently pour the plaster into each mold, avoiding making bubbles. Take care to fill each mold to the top. Gently bang the bottom of each silicone mold against the table to release any trapped air bubbles. The four faces of the wooden molds can be rapped with the edge of a ruler for the same purpose. Allow the plaster to set hard before attempting to remove the casts from the mold. Setting involves an exothermic reaction and the cubes will become quite warm, especially the larger ones.

Once the cubes have been removed from their molds they should be left to dry. This will take several days for the largest ones because of their small area-to-volume ratios. You could compare the rate of drying of the different sized cubes to demonstrate how area-to-volume ratios decrease with increasing size. When each cube is completely dry, as shown by its constant mass, it can be varnished. This step will help protect the surface from scratching and soiling and the corners from chipping. The smallest cubes are best left unvarnished, for two reasons. First, they will not be subjected to the wear and tear of the large cubes. Second, since they have a relatively large area-to-volume ratio, adding varnish will add a disproportionate amount to their mass.

Nuts and washers

I chose nuts and washers for convenience, but any solid objects can be used, provided they are of the same shape and material. Other suggestions include nails, wood screws, bolts, screwdrivers, and ball bearings.

Animals (and plants)

If you are going to use gastropod shells they should be relatively thick, otherwise the departures from the predicted results will be much greater. Alternatively, you could use pickled specimens, where the bodies are still within the shell. Other fairly readily available, variable-sized invertebrates include beetles, cockroaches, mussels, and shrimps. You could also try some vegetables. Tomatoes come in a range of sizes, from the small cherry variety to the giant beefsteak ones – all roughly the same shape. If you are selective, you may also be able to pick out a suitable series of similarly shaped potatoes of different sizes. Onions also vary in size, from the pickling variety, available during the fall, to the large Spanish ones.

Among vertebrates, fishes are ideal because they grow continuously throughout life, so growth series can be obtained for individual species.

LABORATORY 2

Equipment

I was fortunate in obtaining the apparatus used for measuring *E* for a length of wire from the Physics Department of my own university. If you can obtain a vernier scale like the one I used (see Fig. 2.9), it would not be difficult to set up a suitable apparatus. One half of the scale is attached to the end of the wire that is to be loaded (Fig.

Figure A.2. A close-up of the vernier scale used in the apparatus for measuring Young's modulus in a piano wire.

LOAD

A.2). The other half attaches to the unloaded wire, the sole function of the latter being to support this scale. The adjacent edges of the two halves of the vernier scale fit together in a sliding union – like tongue-and-groove planks of wood. They are held in contact by a loosely fitting elastic band. The wire has to be firmly bolted into the ceiling. (Our building maintenance department obligingly bolted the wire into a concrete beam above the dropped ceiling.) Care has to be taken not to kink the wire during installation or during subsequent use. A kink in the wire would give an exaggerated value for the extension, and would therefore make your results unusable.

When you add weights to the scale pan, the wire stretches, moving one half of the vernier scale relative to the other. You can then simply read the extension from the scale.

LABORATORY 3

Equipment

The Instron I have access to is an old machine, controlled by four function buttons: STOP, RAISE, LOWER, and RETURN. The STOP button arrests the movement of the cross-head, but the load cell continues measuring the load acting on the test piece, and the pen recorder and paper chart continue operating, too. Pressing the RAISE button causes the cross-head to elevate slowly, increasing the load on the test piece. The LOWER button has the opposite effect, and the RETURN button also lowers the cross-head, but very rapidly. Since it is difficult to control this last function, it is not used when a test piece is in place. You should bear the following points in mind when operating the Instron or similar test equipment:

1. The Instron has the potential to break bodies as well as test pieces, so treat it with respect.
2. The Instron can tear itself apart if it is left running. As soon as the machine has registered a full-scale deflection on the paper, press the STOP button. When operating the machine keep one finger over the STOP button in readiness.

The speed of the cross-head can be adjusted by changing the gearing of the machine. The gears are housed toward the back, with access through a lift-up lid. Incidentally, raising the lid immobilizes the Instron and therefore acts as a safety device. Chances are that the machine you use will already be set for the appropriate speed and will need no further attention. I use a cross-head speed of 1/2 inch per minute, which is satisfactory. Access to the gears also provides a way of moving the cross-head manually; simply attach a round handle to one of the gear spindles. Hand-cranking is used to make fine adjustments in the position of the cross-head so that the force on the load cell can be brought to zero.

Following is the procedure for conducting tensile tests.

1. Start by carrying out the warm-up and zeroing procedures, as outlined in the protocol for operating the machine.
2. Secure the test piece in the top jaws first. This is because they are free to move, whereas the bottom jaws are fixed. This does not matter very much when testing ductile samples, but brittle materials usually break if attached to the lower jaws first due to the movement caused when you tighten the upper jaws. When removing samples, release the bottom jaws first.
3. Tighten the jaws securely, otherwise the test specimen will slip. However, when using brittle material, too much tightening can crack the sample, so be careful.
4. Clamping the test piece between the jaws will probably exert a small force, activating the load cell and causing it to buzz. This damages the load cell, so you need to remove the force by hand-cranking.
5. Accurately measure the distance between the jaws with calipers or dividers. This measurement will be referred to as the original length of the sample. Record it on the chart paper, together with a small sketch showing the approximate shape of the test piece and the way it is loaded.
6. Set the dial for the appropriate full-scale deflection range, marking this on the chart. Make sure the pen is on the zero line, hand-cranking if necessary until it is.
7. Press the RAISE button, getting ready to hit STOP when the pen reaches the top of the scale (full-scale deflection).
8. Immediately after reading a full-scale deflection, press LOWER, then STOP when the pen reaches the zero line. Hand crank, if necessary, to return to zero.
9. If testing to failure, set the dial for the next (higher) full-scale deflection range and repeat steps 7 and 8.

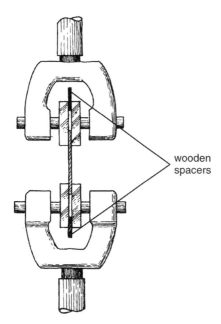

wooden
spacers

Figure A.3. Wooden spacers, clamped in the top half of the jaws of the Instron, reduce the chances of cracking brittle bone samples.

10. Repeat the previous step until the test piece breaks. Be prepared for a fairly loud bang.

Tensile bone samples are easily broken when clamping them in the jaws. It is helpful to clamp a piece of wooden tongue depressor in the top half of the jaws while securing the shoulder region in the lower half (Fig. A.3). This will keep the contact surfaces of the jaws parallel, reducing the risk of cracking and improving the grip. Use the minimum force when tightening the jaws. If the sample starts slipping during the test, the jaws can be tightened some more. In this event, remember to remeasure the distance between the jaws before resuming the test.

Tendon is notoriously difficult to clamp in the jaws. No matter how hard one tries to tighten them, the tendon always seems to slip out. One satisfactory solution used by researchers is to freeze the tendon sample in the clamped jaws. As this is inconvenient, I have tried wrapping each end of the tendon around a piece of wood (a tongue depressor) before clamping. This works quite well.

Compression Testing

By using the cagelike apparatus mentioned in Chapter 3 (Fig. 3.4), you can convert the tensile effort of the Instron into a compressive one. When using this apparatus bear the following points in mind:

1. Fingers have a low compressive strength and can be permanently strained. To ensure against accidents, keep the gear box lid raised, thereby immobilizing the Instron, whenever you are placing samples in the cage.
2. Make sure to position the sample right in the middle of the cage.
3. To compress a sample, just press RAISE in the usual manner, with one finger poised over STOP.

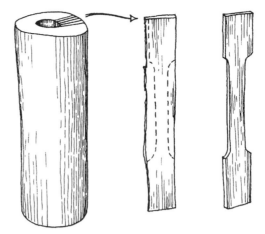

Figure A.4. The steps involved in cutting tensile test pieces from the cannon bone of a horse.

Test Materials

Metal samples for tensile testing are usually available at the facility housing the Instron. They are cut out using a die and conform to a standard pattern. The ones I used were 155 mm long, 12.5 mm wide at the shoulder, and 7 mm wide in the narrow section. Some were cut from an old filing cabinet and had an enamel finish on one side. This was fortuitous because the enamel acted as a stress-coat, cracking under tension to reveal the oblique shearing that occurs. Plexiglas samples will not be available and will have to be cut from a sheet. These need to be much wider than the metal test pieces. The dimensions I found satisfactory were 122 mm long, 25 mm wide at the shoulder, and 11 mm wide in the narrow section. I cut these from a 3 mm thick sheet of Plexiglas. Start by cutting strips 25 mm wide and 122 long, using a circular table saw (ideally, you should enlist the help of an experienced carpenter). Mark each strip for cutting to the final shape, either with a felt-tipped pen or by using masking tape and pen. The approximate shape can be cut out mechanically using a band saw or manually using a hacksaw. Take care not to nick the corners of the shoulders. Finish off by clamping in a vise, using a draw file to straighten up the edges.

I did not attempt to make wooden test pieces for tensile testing. Instead, I simply used wooden tongue depressors, available from surgical suppliers.

The most difficult tensile test pieces to engineer are those made of bone. Compact bone is required, and I used the cannon bone of a horse as a source. For convenience of handling, the bone should be dry, and it should be cut on a band saw. If a power saw is unavailable, a hacksaw can be used. To economize on material and time, I use much smaller test pieces than those used for metals. The following dimensions worked well: length 55 mm, width at shoulders 8 mm, width of narrow section 4 mm. Begin by cutting the cannon bone into short sections, each about 6 cm long. One section will provide several samples, and can be cut into slabs, each about 4 mm thick (Fig. A.4). Discard the slices that are too narrow or that do not have parallel faces. Mark the final shape of the test piece on each slab with pencil or felt-tip pen. Use a powered grindstone for the initial shaping. Use a small file for the final finishing and for removing any nicks. It is important that there be no nicks in the transition from the shoulder to the narrow midsection, otherwise the stress concentration will cause the test piece to break at this point, instead of in the narrow section.

The cubic samples used for compression testing require less work than the tensile test pieces. For wood and Plexiglas, I cut the cubes with an edge of about 10 mm. They do not need to be cubes, and can be more oblong, but their sides have to be parallel. The grain of the wood will probably be obvious, but if not, the direction should be marked on the test piece. The simplest way is to put a dot at either end with a felt-tipped pen. The wooden samples are small enough to be crushed when the Instron is set for the maximum load, but the Plexiglas ones are too large. Therefore make another set of cubes for testing to failure, these with an edge of 3–4 mm. The reason for having two sets is that the smallest cubes are more difficult to handle (forceps are required), and it is therefore more convenient to do most of the testing with the large ones. The bone sample should be cut (from the cannon bone) to a cube of edge 3–4 mm. Since the grain is invisible, it is important to mark the orientation on the sample by reference to the longitudinal axis of the cannon bone. You can do the final finishing of the cubes on a flat sheet of fine sandpaper, making sure not to round the faces.

The tendon samples need to be fairly long (20 cm and more) to get a good grip in the jaws, and they also have to be fairly thick (1 cm or more). I used the long flexor tendons from the leg of a horse. You can probably get these from an abattoir, most likely as partial legs that you will then have to dissect. I have found that tendon keeps well for several days in the refrigerator – sealed securely in freezer bags – and can be stored for extended periods in the freezer.

LABORATORY 4

Equipment

The apparatus shown in Figure 4.6 requires some machining, but it should be a simple task for anyone who works with metal.

A silicone mold is also required to make the fiberglass test pieces. You can make this mold the same way as you did the molds for the small plaster cubes (Laboratory 1). Begin by cutting a wooden model, with dimensions of 75 mm × 25 mm × 6 mm. Treat with oil, as before, prior to coating with silicone sealant. You should make several molds so the class can manufacture several test pieces at the same time.

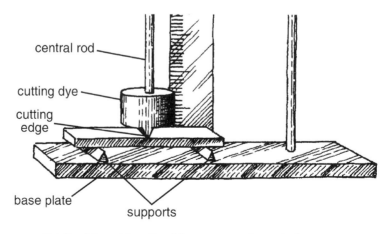

Details of the cutting die of the apparatus shown in Figure 4.6.

LABORATORY 5

Equipment

Ideally, you should use plate glass for grinding and polishing bone sections, as it is thick and therefore unlikely to be broken. The edges should be polished, to avoid cuts. Alternatively, you can seal the edges with masking or electrical tape. Two sheets are required for each set of apparatus (for the two grades of carborundum powder), and these should be about 25 cm square. Plate glass can be purchased and cut to size by glass suppliers, just check your Yellow Pages. Window glass can be used if necessary, obtained from most hardware stores.

Technique

It takes a great deal of practice to prepare good bone slides by hand. The greatest difficulty is in obtaining a uniform thickness, which requires applying the same pressure throughout. The other problem comes from knowing when to stop grinding and start polishing. If you go too far you will finish with no bone. If you do not go far enough, the section will be too thick. Check progress by periodically holding the slide up to the light – it needs to be thin enough to be translucent. When you are getting close, inspect the section under the microscope. Is it thin enough for you to see the bone structure? Once you reach that stage, wash the slide thoroughly under a running tap and start polishing. This last stage is to remove the scratch marks from the grinding stage. Do not polish too much otherwise you run the risk of removing too much bone. Check your progress under the microscope, drying off the section with a tissue after washing and before examining. Although sectioning is a skilled job, your students should be able to make satisfactory enough sections to see the fine structure of bone.

LABORATORY 6

Equipment

Photoelasticity

As noted elsewhere, the best results are obtained using polycarbonate plastic, traded under several different names including Tuffak (manufactured by Rohm and Haas) and Lexan (manufactured by General Electric). A number of different types of Tuffak are available in different thicknesses, some clear, some tinted, some with frosted surfaces. All work well, but some, like Tuffak C, give more discrete contour lines. I have found the people at Rohm and Haas to be very helpful, and if you contact them or one of their retailers, you may be able to obtain some samples; these will be more than enough for your class. Polycarbonate plastic is also available in film, and you can have hours of fun producing psychedelic colors with various setups. Remember that you need some sort of frosting on the side of the plastic, closest to the observer, to enhance the contour lines. The easiest way is to tape the frosted variety of Scotch tape to the surface. Alternatively, you can Scotch tape a piece of translucent acetate film to the plastic. If you use Tuffak with a frosted finish, none of this is necessary.

The polarizing film comes in sheets and can be obtained from educational scientific suppliers like Edmund Scientific Company (101 East Gloucester Pike, Barrington, N.J. 08007-1380). You can use the film in small pieces or sandwich larger sheets between sheets of glass or Plexiglas. A useful variation, for use in class demonstrations, is to mount a piece of the film in a 35 mm transparency holder and use it in a projector to generate a source of polarized light. You can also mount pieces of film in the frames of old spectacles or Scotch tape them to the lenses of functional glasses, taking care to keep the orientation of the film the same on both sides. Polarized sunglasses can also be used as one of the polarizers.

There are all manner of clear plastic products – cups, spoons, shower curtain rings, boxes, lids – that look very attractive when examined through crossed polarizers. The permanent stress contours were formed during manufacture, and some hint of the process may be revealed by the patterns, as when the article was extruded or bent while it was hot. Try bending one of the products and see the formation of stress contours. You may be able to find some plastic item that has been broken and then see the stress patterns at the site of the damage – the possibilities are unlimited.

Specimens

Large robust bones, like the femur of a cow or a horse, can be sectioned using an electric band saw. Bird bones are too delicate and should be sectioned using a small diamond saw attached to a dental engine or electric rotary hand tool. These diamond saws (diamond chips bonded to a metal disc) are used in dentistry and can be purchased from dental suppliers. Try to get one that is about 2 cm in diameter.

LABORATORY 7

Specimens

If you do not have access to a Recent osteology collection, you will need to obtain some skeletons or prepare them for yourself. If you contact a veterinary college, chances are you will be able to find someone who will be prepared to "moonlight" a skeleton for you. Otherwise you will probably be able to obtain a carcass for yourself. Skeletonizing a dog or cat is not a great deal of work, but a cow or a horse is another matter! Aside from the amount of time required to deflesh the bones and boil them up, there is an enormous quantity of soft tissue to dispose of.

Turkey skeletons are easily prepared, especially following Thanksgiving.

LABORATORY 8

Equipment

Sets of equal-volume solid models, ideal for the solid friction experiments, can be obtained from some educational scientific suppliers. The company that used to supply Edmund Scientific is: Halix Electronics, 24 Industrial Blvd., Medford, New York, 11763. The models are called "Geometric Demonstration Models", model number A6202. Towing points can be made by breaking a paper clip in half, bending this into a right angle, and then gluing it to the model with Krazy Glue or

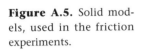
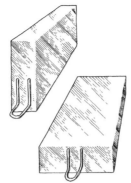

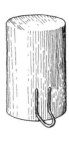

Figure A.5. Solid models, used in the friction experiments.

five-minute epoxy (Fig. A.5). As an alternative, I asked one of our technicians to mold and cast several sets of the plastic models in plaster, embedding the paper clip towing point during the pouring process.

Pigs' Trotters Stew

There is not a great deal of meat on pigs' trotters, but the purpose of including them in a stew is for their collagen. This is obtained from the skin, the tendons, and the ligaments, and, to a lesser extent, the bone. The process is a long, slow one that is best conducted in a Crock-Pot. If you do not have one, use a covered casserole and leave it in a low oven for several hours.

I never bother with precise quantities when making a stew. Simply place some pigs' trotters – four or five should do – in the container. Add two cans of consommé, two large chopped onions, a couple of diced carrots, and about two tablespoons of barley. Sprinkle liberally with mixed herbs and rosemary, and cook for several hours. If using a Crock-Pot, leave it on all day (or overnight) at the lowest setting. If using a casserole in the oven, simmer for about four hours, checking periodically. When it's ready, the bones will be completely free of soft tissue, and the skin will be so soft that it will break between your fingers when pinched. There should be little or no fat, but if there is you can skim it off the top with a spoon. (Alternatively, you can wait until it is cool, finish off the cooling in the refrigerator, and then simply scoop up the solid fat.)

Remove all the bones, using a slotted spoon. Some of them are really small and you may not be able to find them all, so proceed with caution when eating the stew. Give a good stir before serving. Serve with mashed potatoes.

LABORATORY 9

Equipment

The tongue depressors used in the lever experiments should be drilled at 10 mm intervals with holes just large enough to take a paper clip. The pins used to attach the lever to the plywood board are the ones with the hourglass-shaped plastic end pieces.

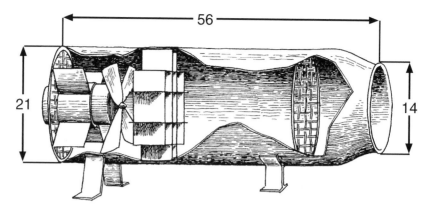

Figure A.6. Construction details of the open wind tunnel. To give some idea of size, I have provided the dimensions of the one I used, in centimeters.

LABORATORY 11

If your university department does not have wind tunnels or the balance and streamlined models that go with them, you might be able to borrow them from a department or college that offers aeronautical engineering classes. Alternatively, you may be able to purchase a set of suitable equipment from an educational scientific supplier. As a last resort, you may be able to have the equipment built in your own department. To assist in the construction, I have given drawings here of the wind tunnel and balance that I used, along with measurements (Fig. A.6). The various models used in these experiments are quite easy to make. For example, I made an airfoil section from a solid piece of Plexiglas for demonstrating the distribution of air pressure over the surface (Chapter 12, Experiments 3–5). I also built its manometer, again using Plexiglas, as detailed later.

Laboratory 11, Part 1, Experiment 1

The disc should be about 35 mm in diameter and only about 2 mm thick. It can be machined from a steel rod of appropriate diameter and welded to a thin metal rod. Alternatively, you can make it by pouring polyester resin into a suitable mold, with a thin rod already in place. I made a disc by cutting a 5 mm slice from a 1 1/2-inch diameter rod of Plexiglas. I then mounted the Plexiglas slice on a 3 mm diameter steel rod.

Laboratory 11, Part 1, Experiments 2 and 3

The sphere and streamlined models can be made by modeling them in plasticine, making molds, and pouring in polyester resin (Fig. A.7). You should position a suitable mounting rod of Plexiglas or a steel rod, about 3 mm in diameter, in the mold prior to pouring. Both models should have the same diameter, which should also be the same as that of the disc in the previous experiment, so that it can be used again here.

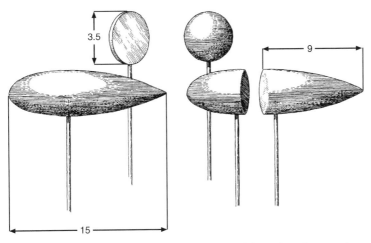

Figure A.7. These models can be built in modeling clay, then used to make molds, making casts in polyester resin. Approximate dimensions are given in centimeters.

Laboratory 11, Part 1, Experiment 5

The different sized discs are cut from a sheet of Fome cor. A short length (15–20 cm) of rod is driven through the disc, parallel with the surface, for attachment – I used a thin welding rod, but a piece of metal coat hanger would work well, too.

Laboratory 11, Part 2, Experiment 4

You need to make a simple recovery device for lifting the models off the bottom of the glycerine-filled glass cylinder at the end of each trial. To do so, bend a length of wire into a loop and attach a piece of netting or gauze over the loop (Fig. A.8).

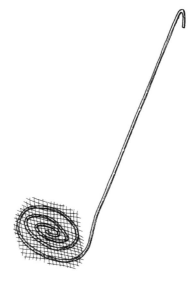

Figure A.8. A simple device for recovering models from tall cylinders of liquid.

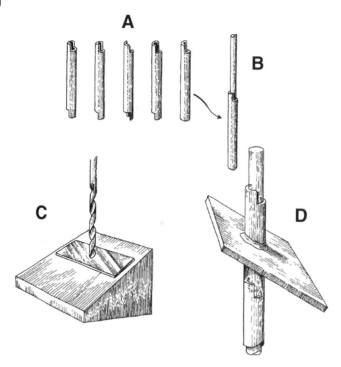

Figure A.9. The manufacture of the inclined plates. A plastic tube is cut into five 2-inch lengths (A). Four of these are step-filed at either end, and the fifth is step-filed at one end. The latter is slipped onto a 12-inch length of plastic rod and cemented at one end (B). The wooden template is used for drilling the hole in the middle of the plate (C). The finished plate has the centrally placed step-filed tube cemented in place (D).

Laboratory 11, Part 3, Experiment 2

To make the set of inclined plates, cut out two square plates (2 × 2 inches) and two rectangular ones (4 × 1 inch) from a sheet of 1/8-inch thick Plexiglas. (Note that English units are used here because the Plexiglas materials used were supplied in these units.) Mark the center of each plate. Using a sheet of fine sandpaper, smoothly taper the two long edges of one of the rectangular pieces of Plexiglas.

Take a piece of Plexiglas tubing with outside and inside diameters of 5/16 and 3/16 inch, respectively. Cut into four 2-inch-long sections. The ends of the tubes are step-filed, as shown in Figure A.9A, so that one-half is about 1/8 inch lower than the other. Cut a fifth 2-inch section, but step-file only one end. Slip the latter onto the end of a 12-inch-long piece of 3/16-inch diameter Plexiglas rod, with the stepped end uppermost. Using a syringe, deliver a drop or two of ethylene dichloride to the joint between the rod and tube. This will spread by capillarity, welding the two pieces of plastic together (Fig. A.9B).

Cut a piece of wood, about 12 inches long, 2 inches wide, and 1 inch deep. Sand or plane down one of its faces such that it is angled at about 20° to the horizontal. This will serve as a template for making the same angle of attack in three of the inclined plates (Fig. A.9C). Cut a second piece of wood, about 3 inches long, 2 inches wide, and 1 inch deep. This is for the square plate with the zero angle of attack. Place one of the square plates on this piece of wood and drill a 5/16-inch hole through its center. Make sure that the hole is vertical to the surface and goes right through the wood. Slip one of the 2-inch lengths of Plexiglas tube

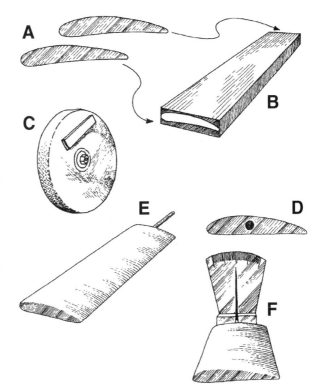

Figure A.10. The manufacture of an airfoil model. First, make metal templates by marking an airfoil shape onto a piece of metal and grinding it down to the outline on a grindstone. Glue the templates (A) to either end of a block of wood or Plexiglas (B). Grind down the block on a grindstone, using the end plates as guides (C). Once the correct shape has been achieved, drill a hole at one end (D) and fit a mounting rod (E). Attach a pointer to the mounting rod and make a scale, marking off degrees using a protractor (F).

through the hole in the square plate and through the hole in the wood, adjusting so that it is at the halfway point of the tube. Cement in place with five-minute epoxy resin and leave to set. Once the epoxy has cured, the plate can be removed from the wooden block and the other side cemented, as before (Fig. A.9D).

Repeat this process for each of the other plates. The reason for drilling them while they are inclined on the wooden block is to ensure that the hole is appropriately angled to the surface. Cement as before. When the four plates have set, place each one, in turn, on the rod and check that the stepped end of the tube interdigitates correctly with the stepped end-stop of the rod. Some small adjustments with a file will probably be necessary. The plates should also interdigitate with one another so that they can be used in tandem, as outlined in the experimental procedures.

LABORATORY 12

Laboratory 12, Experiments 1 and 2

If you do not have an airfoil model, you can easily make one from a piece of wood or Plexiglas, as illustrated (Fig. A.10). When you have completed shaping the wooden model you should sand it with fine sandpaper and seal it with varnish. Plexiglas models should be buffed on a cloth wheel with jewelers' rouge or an equivalent polishing compound.

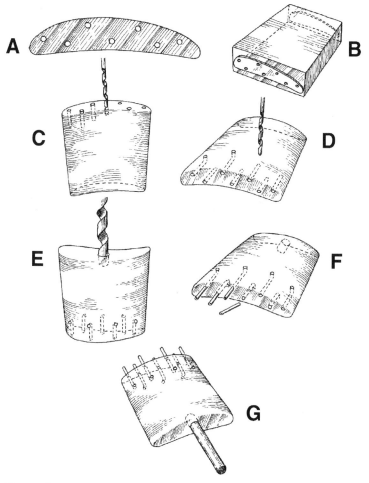

Figure A.11. Steps in modeling a wing section with holes drilled in the top and bottom surfaces: (A) one of the pair of metal templates is drilled with seven holes; (B) the templates are glued to either end of the Plexiglas block; (C) seven holes are drilled into one end of the Plexiglas block; (D) vertical holes are drilled down to the horizontal channels; (E) the other end of the Plexiglas block is drilled with a 3/8-inch hole; (F) 3/4-inch-long sections of Plexiglas tubing are inserted into the small holes and cemented in place; (G) a Plexiglas rod is inserted into the 3/8-inch hole and cemented in place.

Laboratory 12, Experiments 3–5

The airfoil section, and its set of manometers, can be made from Plexiglas with surprising ease. The following is an account of the method I used, including dimensions.

Airfoil

Two airfoil templates, each 3 1/2 inches long, were cut from a 1/8-inch-thick sheet of steel, as shown. One of them was drilled with 1/16-inch holes (Fig. A.11A). A slab of 3/4-inch Plexiglas was cut into a rectangle, 3 1/2 inches wide and 4 inches long. The templates were cemented to each of the 3 1/2-inch ends (Fig. A.11B) using clear acrylic glue (Uhu). When firmly set, the slab was ground down to an airfoil shape on an electric grindstone, using the metal templates as guides.

With the drilled template uppermost, the airfoil section was held vertically in a vise on a drill press (it is important that the airfoil section *is* vertical). Using a 1/16-inch bit, I drilled each of the holes in the template to a depth of 1 1/4 inches (Fig. A.11C). Both templates were then pried off, and each of the ten holes was drilled again, this time with a 3/16-inch bit, but only to a depth of 1/8 inch.

The 3/16-inch drill bit was replaced by a 1 mm bit, and the airfoil laid flat, convex (top) surface uppermost. A strip of 1/2-inch masking tape was laid down across the top surface of the airfoil so that the edge closest to the drilled end was level with the ends of the drill holes. Five pencil lines were drawn across the masking tape, in line with the uppermost set of five holes. These marks served as guides for drilling a set of vertical holes from the surface to connect with the horizontal ones. The drill bit was carefully lowered until it almost touched the top surface, immediately above the first horizontal drill hole (Fig. A.11D). When I was satisfied that the bit was properly lined up, I lowered the drill, drilling down to the horizontal drill hole. The process was repeated for the other four upper horizontal drill holes. After inspection to make sure the vertical holes were correctly positioned, I replaced the 1 mm drill bit with a 1/16-inch one, and the five holes were drilled again to their final size. The reason for initially drilling the vertical holes with the finest drill bit is to minimize damage in the event of a horizontal hole being missed. The incorrect hole can be filled with cement (made by dissolving Plexiglas chips in ethylene dichloride) and drilled again.

The entire procedure was repeated for the lower surface. A 3/8-inch hole was then drilled to a depth of 1/2 inch on the opposite end, for attachment of the supporting rod (Fig. A.11E). Having completed all the machining, I polished the airfoil on a buffing wheel using jewelers' rouge.

Ten 3/4-inch-long sections of 3/16-inch (outside diameter) Plexiglas tube were cut and driven into position in the 1/8-inch-deep holes at the drilled end of the airfoil. These were cemented in place using ethylene dichloride (Fig. A.11F). A 6-inch length of 3/8-inch-diameter Plexiglas rod was inserted into the hole at the other end and cemented with the solvent, thereby finishing the airfoil section (Fig. A.11G). A vertical pointer can be attached to the supporting rod, backed by a suitable scale marked in 2° intervals for measuring the angle of attack.

Manometer Set

A rectangle, measuring 4 × 1 1/2 inches, was cut from a 3/4-inch-thick slab of Plexiglas. A 3/16-inch hole was drilled along the length of the center of the rectangle, as shown, stopping just short of one end (Fig. A.12A). A series of seven vertical holes, centered 1/2 inch apart, were drilled down to the horizontal channel (Fig. A.12B). Seven 6 1/2-inch lengths of 3/16-inch (outside diameter) Plexiglas tubing were cut. These were then located into the seven vertical holes and pushed down to a depth of about 1/4 inch (Fig A.12C). One 3/4-inch section of 3/16-inch tubing was also cut and located into the hole at the end of the rectangular slab. All of the tubes were then cemented in place using ethylene dichloride. One 10 cc

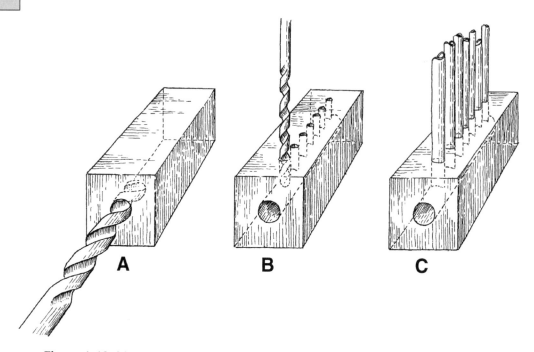

Figure A.12. Main steps in making the manometer set: (A) the Plexiglas block is drilled with a hole that stops just short of the far end; (B) vertical holes are drilled down the horizontal channel; (C) the seven Plexiglas tubes are inserted into the vertical holes and cemented in place.

plastic syringe was taken, and the plunger discarded. Its nozzle was then connected to the short horizontal tube with rubber tubing. The syringe was held vertically in a retort clamp, and colored liquid added. By adjusting the vertical height of the syringe, and the amount of liquid within it, I brought the liquid levels in the seven manometers to the same midpoint level in the tubes.

References

Alexander, R. McN. 1976. Estimates of speeds of dinosaurs. *Nature*, 261:129–130.
  1977. Allometry of the limbs of antelopes (Bovidae). *Journal of Zoology, London*, 183:125–146.
  1984a. Optimum strength for bones liable to fatigue and accidental fracture. *Journal of Theoretical Biology*, 109:621–636.
  1984b. Walking and running. *American Scientist*, 72:348–354.
  1984c. Elastic energy stores in running vertebrates. *American Zoologist*, 24:85–94.
  1989. *Dynamics of Dinosaurs and Other Extinct Giants*. New York: Columbia University Press.
Alexander, R. McN., M. B. Bennett, and R. F. Ker. 1986. Mechanical properties and function of the paw pads of some mammals. *Journal of Zoology, London*, 209:405–419.
Alexander, R. McN., N. J. Dimery, and R. F. Ker. 1985. Elastic structures in the back and their rôle in galloping in some mammals. *Journal of Zoology, London*, 207:467–482.
Alexander, R. McN., and A. S. Jayes. 1983. A dynamic similarity hypothesis for the gaits of quadrupedal mammals. *Journal of Zoology, London*, 201:135–152.
Alexander, R. McN., A. S. Jayes, G. M. O. Maloiy, and E. M. Wathuta. 1979. Allometry of the limb bones of mammals from shrews (*Sorex*) to elephant (*Loxodonta*). *Journal of Zoology, London*, 189:305–314.
Alexander, R. McN., V. A. Langman, and A. S. Jayes. 1977. Fast locomotion of some African ungulates. *Journal of Zoology, London*, 183:291–300.
Alexander, R. McN., G. M. O. Maloiy, B. Hunter, A. S. Jayes, and J. Nturibi. 1979. Mechanical stresses in fast locomotion of buffalo (*Syncerus caffer*) and elephant (*Loxodonta africana*). *Journal of Zoology, London*, 189:135–144.
Auffenberg, W. 1987. Social and feeding behavior in *Varanus komodoensis*. In *Behavior and Neurobiology of Lizards*, eds. N. Greenberg and P. D. MacLean, pp. 301–331. Maryland: National Institute of Mental Health.
Barnett, C. H., and A. Fl. Cobbold. 1962. Lubrication within living joints. *Journal of Bone and Joint Surgery*, 44-B:662–674.
Baudinette, R. V., and K. Schmidt-Nielsen. 1974. Energy cost of gliding flight in herring gulls. *Nature*, 248:83–84.
Beckett, D. 1980. *Brunel's Britain*. Newton Abbot: David and Charles.
Bennet-Clark, H. C., and E. C. A. Lucey. 1967. The jump of the flea: A study of the energetics and a model of the mechanism. *Journal of Experimental Biology*, 47:59–76.
Bennett, A. F. 1991. The evolution of activity capacity. *Journal of Experimental Biology*, 160:1–23.

Bennett, A. F., and J. A. Ruben. 1979. Endothermy and activity in vertebrates. *Science*, 206:649–654.

Bennett, A. F., R. S. Seymour, D. F. Bradford, and G. J. W. Webb. 1985. Mass-dependence of anaerobic metabolism and acid-base disturbance during activity in the salt-water crocodile, *Crocodylus porosus*. *Journal of Experimental Biology*, 118:161–171.

Bennett, M. B., R. F. Ker, N. J. Dimery, and R. McN. Alexander. 1986. Mechanical properties of various mammalian tendons. *Journal of Zoology, London*, 209:537–548.

Biewener, A. A. 1983a. Locomotory stresses in the limb bones of two small mammals: The ground squirrel and chipmunk. *Journal of Experimental Biology*, 103:131–154.

1983b. Allometry of quadrupedal locomotion: The scaling of duty factor, bone curvature and limb orientation to body size. *Journal of Experimental Biology*, 105:147–171.

Biewener, A. A., J. Thomason, A. Goodship, and L. E. Lanyon. 1983. Bone stress in the horse forelimb during locomotion at different gaits: A comparison of two experimental methods. *Journal of Biomechanics*, 16:565–576.

Block, B. A., and D. Booth. 1992. Direct measurement of swimming speeds and depth of blue marlin. *Journal of Experimental Biology*, 166:267–284.

Brown, R. E., and M. R. Fedde. 1993. Airflow sensors in the avian wing. *Journal of Experimental Biology*, 179:13–30.

Burstein, A. H., D. T. Reilly, and V. H. Frankel. 1972. Failure characteristics of bone and bone tissue. In *Perspectives in Biomedical Engineering*, ed. R. M. Kenedi, pp. 131–134. London: Macmillan.

Burstein, A. H., J. D. Currey, V. H. Frankel, and D. T. Reilly. 1972. The ultimate properties of bone tissue: The effects of yielding. *Journal of Biomechanics*, 5:35–44.

Butler, P. J. 1991. Exercise in birds. *Journal of Experimental Biology*, 160:233–262.

Cameron, A. 1981. *Basic Lubrication Theory*. Chichester, England: Ellis Horwood.

Carey, F. G., and B. H. Robinson. 1981. Daily patterns in the activities of swordfish, *Xiphias gladius*, observed by acoustic telemetry. *Fisheries Bulletin*, 79:277–292.

Carrier, D. R. 1987. The evolution of locomotor stamina in tetrapods: Circumventing a mechanical constraint. *Paleobiology*, 13:326–341.

Clancy, L. J. 1975. *Aerodynamics*. London: Pitman.

Colbert, E. H. 1962. *Dinosaurs: Their Discovery and Their World*. London: Hutchinson.

Crick, F. H. C., and J. D. Watson. 1954. The complementary structure of deoxyribonucleic acid. *Proceedings of the Royal Society, A*, 223:80–96.

Currey, J. D. 1984. *The Mechanical Adaptations of Bones*. Princeton: Princeton University Press.

Currey, J. D., and C. M. Pond. 1989. Mechanical properties of very young bone in the axis deer (*Axis axis*) and humans. *Journal of Zoology, London*, 218:59–67.

Demster, D. D. 1958. *The Tale of the Comet*. London: Allan Wingate.

Dewar, H., and J. B. Graham. 1994. Studies of tropical tuna swimming performance in a large water tunnel. I: Energetics. *Journal of Experimental Biology*, 192:13–31.

Dimery, N. J., R. McN. Alexander, and R. F. Ker. 1986. Elastic extension of leg tendons in the locomotion of horses (*Equus caballus*). *Journal of Zoology, London*, 210:415–425.

Drovetski, S. V. 1996. Influence of the trailing-edge notch on flight performance of galliforms. *The Auk,* 113:802–810.

Ellington, C. P. 1991. Limitations on animal flight performance. *Journal of Experimental Biology,* 160:71–91.

Ennos, A. R. 1988. The importance of torsion in the design of insect wings. *Journal of Experimental Biology,* 140:137–160.

Garland, T. 1983. The relation between maximal running speed and body mass in terrestrial mammals. *Journal of Zoology, London,* 199:157–170.

Goldspink, D. F. 1991. Exercise-related changes in protein turnover in mammalian striated muscle. *Journal of Experimental Biology,* 160:127–148.

Gordon, J. E. 1978. *Structures, or Why Things Don't Fall Down.* Harmondsworth, England: Penguin Books.

Gray, J. 1936. Studies in animal locomotion. *Journal of Experimental Biology,* 13:192–199.

Hankin, E. H. 1910. *Animal Flight.* London: Iliffe and Sons.

Heglund, N. C., and C. R. Taylor. 1988. Speed, stride frequency and energy cost per stride: How do they change with body size and gait? *Journal of Experimental Biology,* 138:301–318.

Heglund, N. C., C. R. Taylor, and T. A. McMahon. 1974. Scaling stride frequency and gait to animal size: Mice to horses. *Science,* 186:1112–1113.

Hildebrand, M., and J. P. Hurley. 1985. Energy of the oscillating legs of fast-moving cheetah, pronghorn, jackrabbit, and elephant. *Journal of Morphology,* 184:23–31.

Hoyt, J. W. 1975. Hydrodynamic drag reduction due to fish slimes. In *Swimming and Flying in Nature,* volume 2, eds. T. Y. Wu, C. J. Brokaw, and C. Brennen, pp. 653–672. New York: Plenum Press.

Huxley, J. 1932. *Problems of Relative Growth.* London: Methuen.

Jouventin, P., and H. Weimerskirch. 1990. Satellite tracking of wandering albatross. *Nature,* 343:746–748.

Kaye, G. W. C. , and T. H. Laby. 1995. *Tables of Physical and Chemical Constants,* 16th ed. Harlow, Essex: Longman.

Ker, R. F., M. B. Bennett, S. R. Bibby, R. C. Kester, and R. McN. Alexander. 1987. The spring in the arch of the human foot. *Nature,* 325:147–149.

Kermode, A. C. 1972. *Mechanics of Flight.* London: Pitman Books.

Kosch, B. F. 1990. A revision of the skeletal reconstruction of *Shonisaurus populans* (Reptilia-Ichthyosauria). *Journal of Vertebrate Paleontology,* 10:512–514.

Lachmann, G. V. 1964. Sir Frederick Handley Page. The man and his work. *Journal of the Royal Aeronautical Society,* 68:433–452.

Lanyon, L. E. 1974. Experimental support for the trajectorial theory. *Journal of Bone and Joint Surgery,* 56:160–169.

McGowan, C. 1988. Differential development of the rostrum and mandible of the swordfish (*Xiphias gladius*) during ontogeny and its possible functional significance. *Canadian Journal of Zoology,* 66:496–503.

———. 1989. The ichthyosaurian tailbend: A verification problem facilitated by computed tomography. *Paleobiology,* 15:429–436.

———. 1990. Computed tomography confirms that *Eurhinosaurus* (Reptilia: Ichthyosauria) does have a tailbend. *Canadian Journal of Earth Sciences,* 27:1541–1545.

———. 1991. *Dinosaurs, Spitfires, and Sea Dragons.* Cambridge, Mass.: Harvard.

1992. The ichthyosaurian tail: Sharks do not provide an appropriate analogue. *Palaeontology,* 35:555–570.

McMahon, T. A. 1975. Allometry and biomechanics: Limb bones in adult ungulates. *American Naturalist,* 109:547–563.

Marden, J. H. 1987. Maximum lift production during takeoff in flying animals. *Journal of Experimental Biology,* 130:235–258.

1990. Maximum load-lifting and induced power output of Harris' hawks are general functions of flight muscle mass. *Journal of Experimental Biology,* 149:511–514.

Martin, R. B., and D. B. Burr. 1989. *Structure, Function, and Adaptation of Compact Bone.* New York: Raven Press.

Matthew, P., ed. 1994. *The Guinness Book of Records 1994.* New York: Bantam.

Merriam, J. C. 1908. Triassic Ichthyosauria, with special reference to the American forms. *Memoirs of the University of California,* 1:1–196.

Meyer, G. H. 1867. Die Architektur die Sponiosa. *Archiv für Anatomie, Physiologie und wissenschaftliche Medicine,* 34:615–628.

Meyers, M. J., and K. Steudel. 1985. Effect of limb mass and its distribution on the energetic cost of running. *Journal of Experimental Biology,* 116:363–373.

Motani, R. In press. The first complete forefin of *Grippia longirostris* discovered from an historical specimen. *Palaeontology.*

Motani, R., H. You, and C. McGowan. 1996. Eel-like swimming in the earliest ichthyosaurs. *Nature,* 382:347–348.

Norberg, U. M. 1990. *Vertebrate Flight.* Berlin: Springer-Verlag.

Owen, R. 1840. Note on the dislocation of the tail at a certain point observable in the skeleton of many ichthyosauri. *Transactions of the Geological Society of London,* 5:511–514.

Pascoe, K. J. 1978. *An Introduction to the Properties of Engineering Materials.* London: Van Nostrand Reinhold.

Pennisi, E. 1997. A new view of how leg muscles operate on the run. *Science,* 275:1067

Pennycuick, C. J. 1972. Soaring behavior and performance of some East African birds, observed from a motor-glider. *Ibis,* 114:178–218.

1982. The flight of petrels and albatrosses (Procellariiformes) observed in South Georgia and its vicinity. *Philosophical Transactions of the Royal Society of London,* 300:75–106.

Peters, S. E. 1989. Structure and function in vertebrate skeletal muscle. *American Zoologist,* 29:221–234.

Poore, S. O., A. Sánchez-Haiman, and G. E. Goslow. 1997. Wing upstroke and the evolution of flapping flight. *Nature,* 387:799–802.

Portigliatti-Barbos, M., P. Bianco, A. Ascenzi, and A. Boyde. 1984. Collagen orientation in compact bone: II. Distribution of lamellae in the whole of the human femoral shaft with reference to its mechanical properties. *Metabolic Bone Disease and Related Research,* 5:309–315.

Pritchard, J. J. 1972. General histology of bone. In *The Biochemistry and Physiology of Bone,* Vol. 1, ed. G. H. Bourne. New York: Academic Press, pp. 1–20.

Reilly, D. T., and A. H. Burstein. 1975. The elastic and ultimate properties of compact bone tissue. *Journal of Biomechanics,* 8:393–405.

Reilly, D. T., A. H. Burstein, and V. H. Frankel. 1974. The elastic modulus for bone. *Journal of Biomechanics,* 7:271–275.

Riess, J. 1986. Fortbewegungsweise, Schwimmbiophysik und Phylogenie der Ichthyosaurier. *Palaeontographica A,* 192:93–155.

Riggs, C. M., L. E. Lanyon, and A. Boyde. 1993. Functional associations between collagen fibre orientation and locomotor strain direction in cortical bone of the equine radius. *Anatomy and Embryology,* 187:231–238.

Riggs, C. M., L. C. Vaughan, G. P. Evans, L. E. Lanyon, and A. Boyde. 1993. Mechanical implications of collagen fibre orientation in cortical bone of the equine radius. *Anatomy and Embryology,* 187:239–248.

Roberts, T. J., R. L. Marsh, P. G. Weyand, and C. R. Taylor. 1997. Musculature force in running turkeys: The economy of minimizing work. *Science,* 275:1113–1115.

Ruben, J. A. 1976. Aerobic and anaerobic metabolism during activity in snakes. *Journal of Comparative Physiology,* 109:147–157.

——— 1979. Blood physiology during activity in the snakes *Masticophis flagellum* (Colubridae) and *Crotalus viridis* (Crotalidae). *Comparative Biochemistry and Physiology,* 64:577–580.

Schaller, G. B. 1972. *The Serengeti Lion.* Chicago: University of Chicago Press.

Shapiro, A. H. 1961. *Shape and Flow. The Fluid Dynamics of Drag.* New York: Anchor Books.

Sisson, S. 1953. *The Anatomy of the Domestic Animals.* Philadelphia: W. B. Saunders.

Spector, W. S., ed. 1956. *Handbook of Biological Data.* Philadelphia: W. B. Saunders.

Stinton, D. 1966. *The Anatomy of the Aeroplane.* London: Granada.

Swartz, S. M., M. B. Bennett, and D. R. Carrier. 1992. Wing bone stresses in free flying bats and the evolution of skeletal design for flight. *Nature,* 359:726–729.

Taylor, C. R., A. Shkolnik, R. Dmi'el, D. Baharav, and A. Borut. 1974. Running in cheetahs, gazelles, and goats: Energy cost and limb configuration. *American Journal of Physiology,* 277:848–850.

Thomason, J. J. 1985. The relationship of structure to mechanical function in the third metacarpal bone of the horse, *Equus caballus. Canadian Journal of Zoology,* 63:1420–1428.

——— 1995. To what extent can the mechanical environment of a bone be inferred from its internal architecture? In *Functional Morphology in Vertebrate Paleontology,* ed. J. J. Thomason, pp. 249–263. New York: Cambridge University Press.

Thomason, J. J., A. A. Biewener, and J. E. A. Bertram. 1992. Surface strain on the equine hoof wall *in vivo:* Implications for the material design and functional morphology of the wall. *Journal of Experimental Biology,* 166:145–168.

Torzilli, P. A. 1976. The lubrication of human joints: A review. In *Handbook of Engineering in Medicine and Biology,* eds. D. G. Fleming and B. N. Feinberg, pp. 225–251. Cleveland: CRC Press.

Tucker, V. A. 1993. Gliding birds: Reduction of induced drag by wing-tip slots between the primary feathers. *Journal of Experimental Biology,* 180:285–310.

Wardle, C. S. 1977. Effects of size on the swimming speed of fish. In *Scale Effects in Animal Locomotion,* ed. T. J. Pedley, pp. 299–313. London: Academic Press.

Webb, P. W. 1982. Locomotor patterns in the evolution of actinopterygian fishes. *American Zoolgist,* 22:329–342.

——— 1984. Form and function in fish swimming. *Scientific American,* 251:72–82.

Weihs, D., and P. W. Webb. 1983. Optimization of locomotion. In *Fish Biomechanics,* eds. P. W. Webb and D. Weihs. New York: Praeger Press, 339–371.

Williams, P. L., and R. Warwick. 1980. *Gray's Anatomy.* Edinburgh: Churchill Livingstone.

Wolff, J. D. 1869. Über die Bedeutung der Architektur der spongiosen Substanz. *Zentralblatt für die medizinische Wissenshaft,* 7:223–234.

Wyman, J. 1857. On the cancellate structure of some of the bones of the human body. *Boston Journal of Natural History,* 6:125–140.

Index

DATE DUE
